THE MONTHLY SKY GUIDE

The seventh edition of Ian Ridpath and Wil Tirion's famous guide to the night sky is updated with planet positions and forthcoming eclipses to the end of the year 2011.

The book contains a chapter on the main sights visible in each month of the year and is an easy-to-use companion for anyone wanting to identify prominent stars, constellations, star clusters, nebulae and galaxies, to watch out for meteor showers ('shooting stars'), or to follow the movements of the four brightest planets. Most of the sights described are visible to the naked eye and all are within reach of binoculars or a small telescope. *The Monthly Sky Guide* offers a clear and simple introduction to the skies of the northern hemisphere for beginners of all ages.

IAN RIDPATH is an English astronomy writer and broadcaster, who is also editor of *Norton's Star Atlas* and the *Oxford Dictionary of Astronomy*.

WIL TIRION is a Dutch celestial cartographer, widely regarded as the leading exponent of his art in the world.

THE MONTHLY SKY GUIDE

IAN RIDPATH

illustrated by

WIL TIRION

SEVENTH EDITION

CAMBRIDGE
UNIVERSITY PRESS

CAMBRIDGE UNIVERSITY PRESS
Cambridge, New York, Melbourne, Madrid, Cape Town, Singapore, São Paulo

Cambridge University Press
The Edinburgh Building, Cambridge CB2 2RU, UK

www.cambridge.org
Information on this title: www.cambridge.org/9780521684354

First published 1987
Reprinted 1989
Second edition 1990
Reprinted 1991
Third edition 1993
Reprinted 1994
Fourth edition 1996
Reprinted 1997, 1998
Fifth edition 1999
Reprinted 2001
Sixth edition 2003
Reprinted 2006

Printed in the United Kingdom at the University Press, Cambridge

A catalogue record for this publication is available from the British Library

ISBN-13 978-0-521-68435-4 paperback
ISBN-10 0-521-68435-8 paperback

Contents

For not in vain we watch the constellations,
Their risings and their settings, not in vain
The fourfold seasons of the balanced year.

Teach me to know the paths of the stars in heaven,
The eclipses of the Sun and the Moon's travails

From The Georgics *by Virgil,*
translated into English verse
with an introduction and notes
by L. P. Wilkinson (Penguin Classics, 1982),

Introduction

Stars are scattered across the night sky like sequins on velvet. Over 2000 of them are visible to the unaided eye at any one time under the clearest conditions, but most are faint and insignificant. Only a few hundred stars are bright enough to be prominent to the naked eye, and these are plotted on the monthly sky maps in this book. The brightest stars of all act as signposts to the rest of the sky, as shown on pages 14–15. It is a welcome fact that you need to know only a few dozen stars to find your way around the sky with confidence. This book will introduce you to the stars month by month, without the need for optical aid, so that you become familiar with the sky throughout the year.

What is a star?

All stars are suns, blazing balls of gas like our own Sun, but so far away that they appear as mere points of light in even the most powerful telescopes. At the centre of each star is an immense natural nuclear reactor, which produces the energy that makes the star shine. Stars can shine uninterrupted for billions of years before they finally fade away.

Many bright stars have noticeable colours – for example, Antares, Betelgeuse and Aldebaran are reddish-orange. A star's colour is a guide to its temperature. Contrary to the everyday experience that blue means cold and red is hot, the bluest stars are actually the hottest and the reddest stars are the coolest. Red and orange stars have surfaces that are cooler than that of the Sun, which is yellow-white. The hottest stars of all appear blue-white, notably Rigel, Spica and Vega. On the star charts in this book, the brightest stars are coloured as they appear to the unaided eye. Faint stars show no colour to the eye at all. Star colours are more distinct when viewed through binoculars or telescopes.

The familiar twinkling of stars is nothing to do with the stars themselves. It is caused by currents of air in the Earth's atmosphere, which produce an effect similar to a heat haze. Stars close to the horizon twinkle the most because we see them through the thickest layer of atmosphere (*see diagram*). Bright stars often flash colourfully from red to blue as they twinkle; these colours are due to the star's light being broken up by the atmosphere.

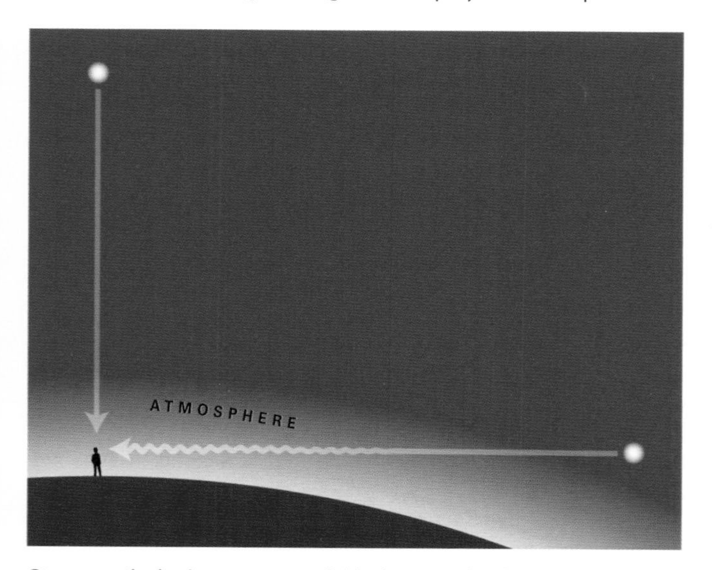

Stars near the horizon seem to twinkle, because their light passes through more of the Earth's atmosphere than light from stars overhead.

What is a planet?

Nine planets, including the Earth, orbit the Sun. The basic difference between a star and a planet is that stars give out their own light but planets do not. Planets shine in the sky because they reflect the light of the Sun. They can consist mostly of rock, like our Earth, or they can be composed of gas and liquid, as are Jupiter and Saturn.

Planets are always on the move, so they cannot be shown on the maps in this book. The three outermost planets – Uranus, Neptune and Pluto – are too faint to be seen with the naked eye. The innermost planet, Mercury, keeps so close to the Sun that it is perpetually engulfed in twilight. So there are only four planets that are prominent to the eye in the night sky: Venus, Mars, Jupiter and Saturn. The positions of these four planets each month for a five-year period are given in the monthly notes in this book. The planets appear close to the plane of the *ecliptic*, the Sun's yearly path around the sky, which is marked as a dashed green line on the maps.

The brightest planet is Venus, for two reasons: it comes closer to the Earth than any other planet, and it is entirely shrouded by highly reflective clouds. Venus is popularly termed the morning or evening 'star', seen shining brilliantly in twilight before the Sun rises or after it has set. As Venus orbits the Sun it goes through phases like those of the Moon, noticeable through small telescopes and binoculars.

The second-brightest planet as seen from Earth is Jupiter, the largest planet of the Solar System. Binoculars reveal its rounded disk and four brightest moons. Mars, when closest to us, appears as a bright red star, but it is too small to show much detail through small telescopes. Saturn at its closest appears to the eye like a bright star, and binoculars just show the outline of the rings that girdle its equator.

It is often said that planets do not twinkle, but this is not entirely correct. Some twinkling of planets can be seen on particularly unsteady nights, but since planets are not point sources they certainly twinkle far less than stars.

What is a constellation?

About 4000 years ago, people of the eastern Mediterranean began to divide the sky into easily recognizable patterns, to which they gave the names of their gods, heroes and fabled animals. Such star patterns are known as constellations. They were useful to seamen for navigation and to farmers who wanted to tell the time of night or the season of the year. By the time of the Greek astronomer Ptolemy in AD 150, 48 constellations were recognized.

Since then, various astronomers have introduced new figures to fill the gaps between the existing ancient constellations. Many of the new groups lie in the far southern part of the sky that was invisible to the Greeks. Some of the newly invented constellations have since been abandoned, others have been amalgamated, and still others have had their names or boundaries changed. This haphazard process has left a total of 88 constellations, of all shapes and sizes, covering the entire sky like pieces of a jigsaw. They all have Latin names. The constellation names and boundaries are laid down by the International Astronomical Union, astronomy's governing body.

The stars in a constellation are usually unrelated, lying at widely differing distances from us and from each other (*see*

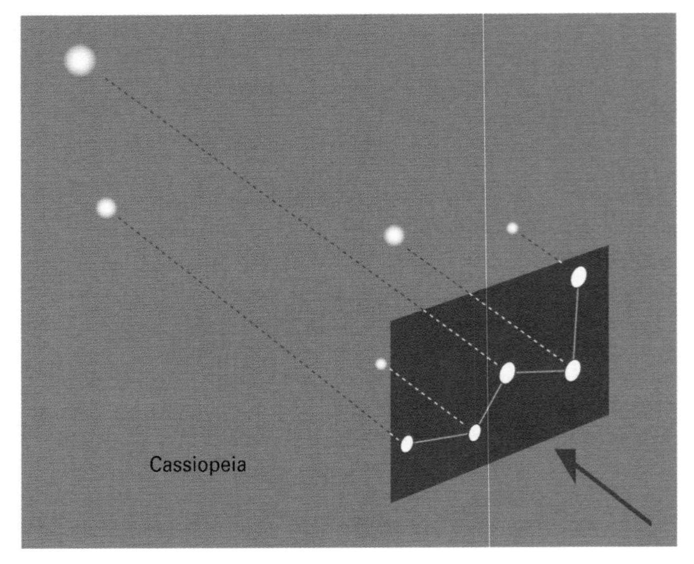

Stars in a constellation lie at different distances from us, as in the case of the five that make up the distinctive W-shape of Cassiopeia.

diagram above), so the patterns they form in the sky are entirely accidental. One fact which dismays beginners is that few constellations bear any resemblance to the objects they are supposed to represent. Constellations are best thought of not as pictures in the sky but as a convenient way of locating celestial objects.

How did the stars get their names?

Most bright stars, and several not-so-bright ones, have strange-sounding names. Other stars are known merely by letters and numbers. These designations arose in various ways, as follows.

A number of star names date back to Greek and Roman times. For example, the name of the brightest star in the sky, Sirius, comes from the Greek word for sparkling or scorching, in reference to its brilliance. The name of another bright star, Antares, is also Greek and can be translated as 'like Mars' or 'rival of Mars', on account of its strong red colour, similar to that of the planet Mars. The brightest star in the constellation Virgo is named Spica, from the Latin meaning 'ear of wheat', which the harvest goddess Virgo is visualized holding.

But most of our star names are Arabic in origin, and were introduced into Europe in the Middle Ages through the Arab conquest of Spain. For example, Aldebaran is Arabic for 'the follower', from the fact that it follows the star cluster known as the Pleiades across the sky. Fomalhaut comes from the Arabic meaning 'mouth of the fish', from its position in the constellation Piscis Austrinus, the Southern Fish. Betelgeuse is a corruption of the Arabic *yad al-jawza*, meaning 'the hand of Orion'. Its name is often mistranslated as 'armpit of the central one'.

In all, several hundred stars have proper names, but only a few dozen names are commonly used by astronomers. Usually, astronomers refer to stars by their Greek letters, assigned in 1603 by the German astronomer Johann Bayer; hence these designations are known as Bayer letters. On this system, Betelgeuse is Alpha (α) Orionis, meaning the star Alpha in the constellation of Orion. Another system of labelling stars is by their numbers in a star catalogue compiled by the English astronomer John Flamsteed. These are known as Flamsteed numbers, and they are applied to fainter stars that do not have Bayer letters, such as 61 Cygni. (Note that the genitive, or possessive, case of a constellation's Latin name is always used with Bayer letters and Flamsteed numbers). Stars too faint to be included in these systems, or stars with particular characteristics, are referred to by the numbers assigned to them in a variety of specialized catalogues.

How far are the stars?

So remote are the stars that their distances are measured not in kilometres or miles but in the time that light takes to travel from them to us. Light has the fastest speed in the Universe, 300,000 km/sec (186,000 mile/sec). It takes just over 1 second to cross the gap from the Moon to the Earth, 8.3 minutes to reach us from the Sun, and 4.4 years to reach the Earth from the nearest star, Alpha Centauri. Hence Alpha Centauri is said to be 4.4 light years away.

A light year is equivalent to 9.5 million million km (5.9 million million miles), so that in everyday units Alpha Centauri is about 40 million million km (25 million million miles) away. Even our fastest space probes would take about 80,000 years to get to Alpha Centauri, so there is no hope of exploring the stars just yet.

Most of the stars visible to the naked eye lie from dozens to hundreds of light years away. It is startling to think that the light entering our eyes at night left those stars so long ago. The most distant stars that can be seen by the naked eye are over 1000 light years away, for example Deneb in the constellation Cygnus and several of the stars in Orion. Only the most luminous stars, those that blaze more brightly than 50,000 Suns, are visible to the naked eye over such vast distances. At the other end of the scale, the feeblest stars emit less than a thousandth of the light of the Sun, and even the closest of them cannot be seen without a telescope.

How bright are the stars?

The stars visible to the naked eye range more than a thousandfold in brightness, from the most brilliant one, Sirius, to those that can only just be glimpsed on the darkest of nights. Astronomers term a star's brightness its *magnitude*. The magnitude system is one of the odder conventions of astronomy.

Naked-eye stars are ranked in six magnitude classes, from first magnitude (the brightest) to sixth magnitude (the faintest). A difference of five magnitudes is defined as equalling a brightness difference of exactly 100 times. Hence a step of one magnitude corresponds to a difference of about 2.5 times in brightness. A difference of two magnitudes corresponds to a brightness difference of $2.5 \times 2.5 = 6.3$ times. Three magnitudes equals a brightness difference of $2.5 \times 2.5 \times 2.5 = 16$ times, and so on.

A star 2.5 times brighter than magnitude 1.0 is said to be of magnitude 0. Objects brighter still are assigned negative magnitudes. Sirius, the brightest star in the entire sky, has a magnitude of −1.4.

The magnitude system can be extended indefinitely to take account of the brightest and the faintest objects. For instance, the Sun has a magnitude of −27. Objects fainter than sixth magnitude are classified in succession as seventh magnitude, eighth magnitude, and so on. The faintest objects that can be detected by telescopes on Earth are about magnitude 26.

What are double stars?

Most stars are not single, as they appear to the naked eye, but have one or more companion stars, a number of which can be seen through small telescopes, and some with binoculars. The visibility of the companion star depends on its brightness and its distance from the primary star – the closer the stars are together, the larger the aperture of telescope that is needed to separate them.

Usually, the members of a star family all lie at the same distance from us and orbit each other like the planets around the Sun. A pair of stars genuinely related in this way is known as a binary. But sometimes one star simply lies in the background of

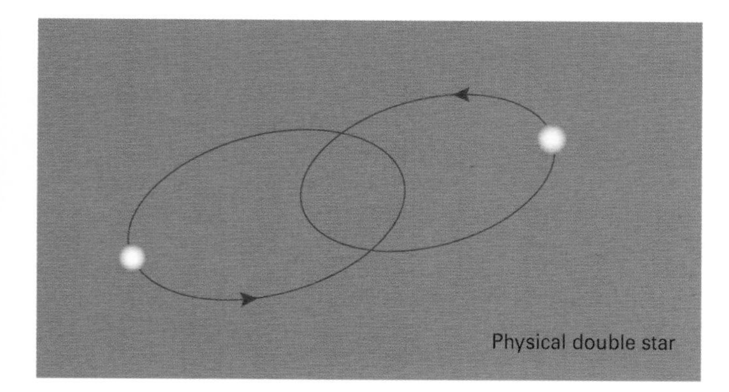

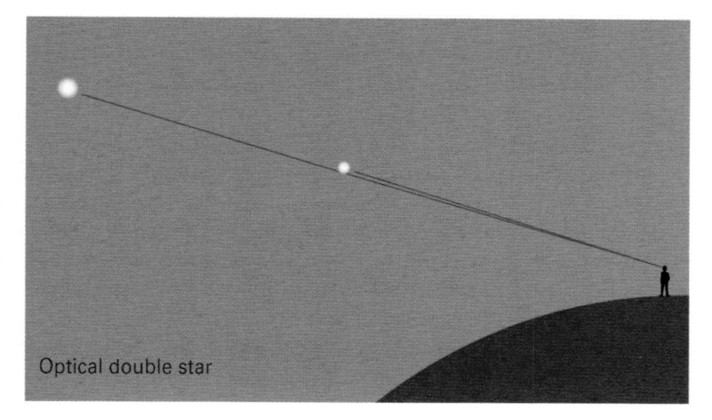

Most double stars are related and orbit each other (a physical double, or binary, top). But in some cases two stars lie by chance in the same line of sight, giving rise to an optical double.

another, at a considerably greater distance from us (*see diagram above*). Such a chance alignment of two unrelated stars is termed an optical double.

One of the attractions of double and multiple stars is the various combinations of star colours that are possible. For example, one of the most beautiful double stars is Albireo in the constellation Cygnus, which consists of amber and green stars, looking like a celestial traffic light (see page 46). Other impressive double and multiple stars are highlighted in the constellation notes.

What are variable stars?

Not all stars are constant in brightness. Certain stars change noticeably in brightness from night to night, and some even from hour to hour. The brightest of the varying stars is Betelgeuse in Orion, which ranges a full magnitude (i.e. 2.5 times) between maximum and minimum intensity, taking many months or even years to go from one peak to the next. The variations in the light output of Betelgeuse are caused by changes in the actual size of the star. Although the variations of Betelgeuse have no set period, a number of stars pulsate regularly, like a beating heart, every few days or weeks. These are known as *Cepheid variables*, and they follow the rule that the longer the cycle of variation, the more luminous is the star.

Some stars which appear to vary in brightness are actually close pairs in which one star periodically eclipses the other, temporarily blocking off its light from us. The prototype of these so-called *eclipsing binaries* is Algol, in the constellation Perseus, which fades to one-third its usual brightness every 2 days 20 hours 49 minutes. This and other interesting variable stars are mentioned in the constellation notes.

The most spectacular of the varying stars are the *novae*, faint stars which can flare to thousands of times their usual brightness for a few days before sinking back into obscurity. Novae got their name, which is Latin for 'new', because in the past they were really thought to be new stars appearing in the sky. Now we know that a nova is actually an old, dim star that has flared up because gas has fallen onto it from a neighbour. Nova outbursts are unpredictable, and are often first spotted by amateur astronomers.

What is a shooting star?

During your sky watching, you will from time to time see a bright streak of light dash across the sky like a cosmic laser beam, lasting no more than a second or so. This is a *meteor*, popularly termed a shooting star. Do not misidentify shooting stars with satellites or high-flying aeroplanes, which look like moving stars but drift at a more leisurely pace.

Despite their name, shooting stars are nothing to do with stars at all. They are particles of dust shed by comets and are usually no bigger than a grain of sand. They plunge into the Earth's atmosphere at speeds from 10 to 75 km/sec (6 to 46 mile/sec). We see a glowing streak of hot gas as the dust particle burns up by friction with the atmosphere about 100 km (60 miles) above the Earth. The brightest meteors outshine the stars, and some break up into glowing fragments. Occasionally, much larger chunks of rock and metal penetrate the atmosphere and land on Earth. These are called meteorites.

On any clear night, several meteors are visible to the naked eye each hour, burning up at random. These are known as *sporadic* meteors. But at certain times of the year the Earth ploughs through dust storms moving along the orbits of comets. We then see a meteor shower, when as many as 100 meteors may be visible each hour.

Owing to an effect of perspective, all the members of a meteor shower appear to diverge from a small area of sky known as the *radiant*. The meteor shower is named after the constellation in which its radiant lies: the Perseids appear to radiate from the constellation of Perseus, the Geminids from Gemini and so on. Meteor showers recur on the same dates each year, although the intensity of a shower can vary from year to year. As the Earth penetrates the swarm of dust, the number of meteors visible each hour builds up over a few days to its maximum, and then falls away again. Amateur astronomers monitor the progress of a shower by counting the number of meteors visible, and estimating their brightnesses.

When watching for meteors from a shower, do not gaze directly at the radiant, but scan about 90° to the side of it, where the meteor trails will be longest. A list of the year's main meteor showers is given in the table overleaf. Note that the maximum number of meteors quoted will be seen only in dark conditions, when the radiant is high in the sky. If the radiant is low, or if the sky is bright (for instance from moonlight), the number of meteors visible will be very much smaller.

What is a comet?

People often confuse comets with shooting stars, but they are quite different things. Whereas a shooting star appears as a brief streak of light, a comet looks like a hazy smudge hanging in the sky. A comet's movement against the star background is noticeable only over a period of hours, or from night to night. Comets are giant snowballs of frozen gas and dust that move on highly elongated orbits around the Sun, taking from a few years to many thousands of years to complete one circuit.

About three dozen comets are seen by astronomers each year. These are a mixture of new discoveries and previously known examples returning to the inner part of the Solar System. Most comets are too faint to be seen without a large telescope. Only rarely, perhaps every ten years or so, does one become bright

Meteor showers

Name of shower	Date of maximum	Number of meteors visible per hour at maximum (approx.)
Quadrantids	January 3–4	100
Lyrids	April 21–22	10
Eta Aquarids	May 5–6	35
Delta Aquarids	July 28–29	20
Perseids	August 12–13	75
Orionids	October 22	25
Taurids	November 4	10
Leonids	November 17–18	10
Geminids	December 13–14	75

enough to be prominent to the naked eye, as Comet Hale–Bopp was in 1997.

A bright comet is awesomely beautiful. From its glowing head, a transparent tail of gas and dust extends for millions of kilometres, always pointing away from the Sun. The dust released by the comet disperses into space, some of it eventually being swept up by the Earth to appear as meteors.

What is an eclipse?

At an eclipse of the Sun, the Moon passes in front of the Sun and cuts off its light from part of the Earth; at an eclipse of the Moon, the Moon enters the Earth's shadow cast in space and hence is darkened. Eclipses occur only occasionally, when Sun, Moon and Earth line up suitably, but at least two solar eclipses are visible somewhere on Earth every year, and most people will see an eclipse of the Moon every couple of years. Solar and lunar eclipses for a five-year period are listed in the monthly notes in this book (note that not every month has an eclipse).

Solar eclipses are visible from only a limited area on Earth. When the Sun is entirely covered by the Moon the eclipse is said to be total, but a total solar eclipse lasts only a few minutes and is rare from any given point on Earth. A partial solar eclipse can last several hours and is visible over a much wider area.

A special case of a solar eclipse is known as an annular eclipse. This occurs when the Moon is at its farthest from Earth and hence its apparent size is not quite sufficient to cover the Sun completely – so instead of a total eclipse, a ring of sunlight is left around the Moon. The name annular eclipse comes from the Latin *annulus*, meaning ring.

By contrast with the limited visibility of solar eclipses, eclipses of the Moon can be seen from anywhere on Earth that the Moon is above the horizon. A total lunar eclipse can last over an hour.

What is the Milky Way?

On clear, dark nights, a hazy band of starlight arches across the heavens. The Greeks called it the Milky Way. We know that the Milky Way consists of innumerable stars comprising an enormous wheel-shaped structure, the Galaxy, of which our Sun is a member. The stars that are scattered randomly over the night sky, forming the constellations, are among the nearest to us in the Galaxy. The more distant stars are concentrated into the crowded band of the Milky Way. The Milky Way, therefore, is the rest of our Galaxy as seen from our position within it.

The Galaxy's centre lies in the direction of the constellation Sagittarius, where the Milky Way star fields are particularly dense. Our Sun lies approximately two-thirds of the way from the hub to the rim of the Galaxy, which is about 100,000 light years in diameter. Beyond the edge of the Galaxy is empty space, dotted with other galaxies.

On the monthly star maps, the Milky Way is indicated in light blue. Sparkling star fields spring into view if you sweep along this region with binoculars or small telescopes.

What is a star cluster?

In places, stars congregate in clusters, some of which are visible to the naked eye, most notably the group called the Pleiades in the constellation Taurus. There are two sorts of star cluster, distinguished by the types of stars in them and their location in the Galaxy. *Open clusters* are loose groupings of young stars dotted along the spiral arms of our Galaxy. Some open clusters are still surrounded by the remains of the gas clouds from which they were born. Open clusters can contain from a handful of stars up to a few thousand stars.

Altogether different in nature are the ball-shaped *globular clusters*, mostly found well away from the plane of the Milky Way. They are swarms of up to 300,000 stars, much more tightly bunched than in open clusters. The stars in globular clusters are very old – indeed, they include some of the oldest stars known. Since globular clusters are much farther from us than open clusters, they appear fainter. The best globular cluster for northern observers is M13 in the constellation Hercules (see page 38).

What is a nebula?

Between the stars are vast clouds of gas and dust known as *nebulae*, the Latin for mist, a word that accurately describes their foggy appearance. They are best seen in clear country skies, away from smoke and streetlights.

Some nebulae shine brightly, whereas others are dark. The most famous bright nebula is in Orion, visible to the naked eye as a softly glowing patch. The Orion Nebula is a gas cloud from which stars are forming, and the new-born stars at its centre light up the surrounding gas (see page 18). Other nebulae remain dark because no stars have yet formed within them. Dark nebulae become visible when they are silhouetted against dense star fields or bright nebulae. The Milky Way star fields in Cygnus are bisected by a major dark nebula, the Cygnus Rift.

Certain nebulae are formed by the deaths of stars, including the so-called *planetary nebulae* which, despite their name, have nothing to do with planets. The name arose because, when seen through small telescopes, they often look like the disks of planets. Actually, planetary nebulae are glowing shells of gas sloughed off by stars like the Sun at the ends of their lives. Stars that are much more massive than the Sun die in violent explosions, splashing fountains of luminous gas into space. The most famous remnant of an exploded star is the Crab Nebula in the constellation Taurus (see page 62).

What is a galaxy?

Some objects that at first sight resemble nebulae are actually distant systems of stars beyond our Milky Way – other galaxies, many millions of light years apart, dotted like islands throughout the Universe. The smallest galaxies consist of approximately a million stars, while the largest galaxies contain a million million stars or more. We live in a fair-sized galaxy of at least 100,000 million stars.

Galaxies come in two main types: spiral and elliptical. Spiral galaxies have arms consisting of stars and gas winding outwards from their star-packed hubs. A sub-species of spiral galaxies, called barred spirals, have a bar of stars across their centre; the

Naming star clusters, nebulae and galaxies

Star clusters, nebulae and galaxies are often referred to by numbers with the prefix M or NGC. The M refers to Charles Messier, an eighteenth-century French comet hunter who compiled a list of over 100 fuzzy-looking objects that might be mistaken for comets. Astronomers still use Messier's designations, and enthusiasts avidly track down the objects on his list. In 1888 a far more comprehensive listing of star clusters and nebulous objects was published, called the *New General Catalogue of Nebulae and Clusters of Stars*, or NGC for short. This was followed by two supplements known as the *Index Catalogues* (IC), bringing the total number of objects catalogued to 13,000, most of them only within reach of large telescopes. In this book we use the Messier numbers where they exist, or NGC and IC numbers otherwise.

spiral arms emerge from the ends of the bar. Our Galaxy is now thought to be a barred spiral. Elliptical galaxies have no spiral arms. They range in shape from almost spherical to flattened lens shapes. Time-exposure photographs with large telescopes are needed to bring out the full beauty of galaxies. Through binoculars and small telescopes, most galaxies appear only as hazy patches of light. Like nebulae, galaxies are best seen in clear, dark skies. The nearest major galaxy to us is just visible to the naked eye as a faint smudgy patch similar in apparent width to the full Moon in the constellation of Andromeda (see page 54).

How can I see satellites?

While casually watching the sky you may well see what looks like a moderately bright star drifting along, taking a few minutes to pass from one horizon to the other. This will be an artificial satellite, orbiting the Earth a few hundred kilometres up and lit by the Sun. (To confirm that it is not a high-flying aircraft, check with binoculars which will reveal an aircraft's lights, whereas a satellite will remain a star-like point.) Dozens of satellites are bright enough to be seen by the naked eye, most notably the International Space Station (ISS), which can be dazzling. You may even see the Space Shuttle on one of its periodic trips to and from the ISS. Satellites are easiest to see on summer nights, because the Sun then never gets far below the horizon and satellites are always in daylight. In winter, satellites will be visible only shortly after sunset and shortly before sunrise. Sometimes a satellite may fade in or out partway across the sky as it emerges from or vanishes into the Earth's shadow.

Some satellites flash as they move, particularly discarded rocket stages which are tumbling in orbit. The most extreme examples of flashing satellites are a system of communications satellites called Iridium. These are normally invisible, but occasionally sunlight reflects off their flat transmitter panels to create a brilliant flare that can last for several seconds.

Satellite predictions for any location can be obtained from an Internet website called Heavens Above:
`http://www.heavens-above.com`

How to look at the stars

To begin stargazing you need nothing more than your own eyes, supplemented by a modest pair of binoculars. The monthly star charts in this book are designed to be used for stargazing with the naked eye. With these charts, you will be able to identify the stars and constellations visible on any night of the year.

Specific constellations are featured in more detail in the monthly star notes, and to study the objects described in them requires some form of optical aid. Optical instruments collect more light than the naked eye does, thus showing fainter objects as well as making objects appear bigger by magnifying them. In astronomy, the ability of an instrument to collect light is often more important than the amount by which it magnifies. This is particularly so in the case of binoculars, which are an indispensable starting instrument for any would-be stargazer.

Binoculars usually have relatively low magnifications of between six and ten times, so they will not show detail on the planets, but their light-grasp will bring into view many stars and nebulae that are beyond reach of the naked eye. Binoculars have a much wider field of view than telescopes, and so are better suited for studying extended objects such as scattered star clusters. Even if you subsequently obtain a telescope, binoculars will always remain of use.

Binoculars bear markings such as 6 × 30, 8 × 40 or 10 × 50. The first figure is the magnification, and the second figure is the aperture of the front lenses in millimetres: the larger the aperture, the greater the light-grasp and hence the fainter the objects that you can see. Binoculars with high magnifications of 12 times or more are occasionally encountered, but these need to be mounted on a tripod to hold them steady.

Telescopes are rather like telephoto lenses, but whereas a telephoto lens on a camera is described by its focal length a telescope is described by its aperture. Remember that a telescope with an aperture of, say, 50 mm has the same light-grasp as a pair of 50-mm binoculars, but it can cost several times more. The main advantage of a telescope over binoculars is that it has higher magnification.

The smallest telescopes – those with apertures up to 75 mm (3 inches) – are of the refracting type, like a spyglass, which you look straight through. Larger telescopes, starting with apertures around 100 mm (4 inches), are usually reflectors, in which mirrors collect the light and bounce it into an eyepiece. Reflectors are cheaper to make in large sizes than refractors, and they usually have shorter tubes, which is more convenient. A third breed of telescope now commonly encountered combines lenses and mirrors. These are termed catadioptric telescopes and are best thought of as a modified form of reflector. Their main advantage is that they are compact and portable, but they are more expensive than basic reflectors, particularly the so-called GOTO variety that has computer controls for finding objects.

A problem that becomes apparent on looking through a telescope is the unsteadiness of the atmosphere. Turbulent air currents make the image of a star or planet seem to boil and jump around, which limits the amount of detail that can be discerned, particularly at higher magnifications. The steadiness of the atmosphere is known as the *seeing*, and it changes markedly from night to night. You will be able to see more detail through a telescope on nights with steady air (i.e. good seeing).

Telescopes have interchangeable eyepieces which offer a range of magnifications. High magnifications are useful when studying fine detail on the planets or separating the close components of a double star. But no matter how much magnification you use on a small telescope, it will not show as much detail as a larger telescope. In fact, too high a power on a telescope will make the image so faint that you will end up seeing less than if you had used a lower power. A good rule of thumb is to keep to a maximum magnification of 20 times per 10 mm of aperture (50 times per inch). Do not be tempted by small telescopes that offer magnifying powers of many hundreds of times. In practice, such high magnifications will be unusable. Most of the objects mentioned in this book are within view of telescopes with an aperture of 100 mm or less, using low to medium magnifying powers.

Finally, be sure to wrap up warm when going out to observe, even on a mild evening. If you are observing with the naked eye or binoculars, sit in a reclining chair or lie on a sunbed. A comfortable observer sees more than an uncomfortable one.

How to use the maps in this book

There are twelve all-sky charts in this book, one for each month of the year, usable every year. They show the sky as it appears between latitudes 30° north and 60° north at fixed times of night that are convenient for observation. To understand the maps, imagine the night sky as a dome overhead. The maps are a flattened representation of that dome. When you look upwards you must think of the maps as being stuck to the dome of the sky, arching over you on all sides.

To use the maps, turn them so that the direction marked at the bottom of the map – north, south, east or west – matches the direction you are facing. The map then depicts the sky as it appears at 10 p.m. in mid-month. The sky will look the same at 11 p.m. at the start of the month and 9 p.m. at the end of the month. A table at the bottom of each monthly map shows the times at which the sky appears the same in other months. Note: one hour must be added to these times when daylight saving ("Summer Time") is in operation, designated DST in the key. For information on changeover times to DST, see
`http://webexhibits.org/daylightsaving/b.html`

If you wish to observe at times other than those in the table, you must make allowance for the changing appearance of the sky caused by the rotation of the Earth, which turns through 15° every hour. When your observing time differs by a full two hours from the time shown on a map, you will have to change maps. For every additional two hours' difference in time, go forwards or backwards another map as necessary. Inspection of the tables at the bottom of each chart will show that, for example, the appearance of the sky at 10 p.m. in mid-May is the same as at 6 a.m. in mid-January.

Changing maps will be necessary to find when planets are visible if they are located in constellations not above the horizon at 10 p.m. in a given month. For example, if a planet lies in Sagittarius during January, we have to move forward four maps to find the map on which Sagittarius first appears above the horizon – the May map. The time table tells us that Sagittarius, and hence the planet, do not rise until nearly dawn in January.

Four horizons are shown on the maps, for observers at latitude 30° north, 40° north, 50° north and 60° north. It should not be difficult to work out where the horizon falls on the maps for your specific latitude. Note how your latitude affects the stars that are visible. You will be able to use the maps from anywhere between latitudes 30° north and 60° north, and there will be little discrepancy between what the maps show and the way the sky appears for up to 10° outside these limits.

The monthly all-sky maps show the brightest stars down to magnitude 5.0, realistically depicting the sky as seen by the naked eye on an averagely clear night. The maps of individual constellations show stars down to magnitude 6.5, plus fainter selected stars and numerous objects of interest for users of binoculars and small telescopes.

Scale in the sky

One difficulty when matching up the maps in a book with the appearance of the real sky is the matter of scale. Most people overestimate the size of objects in the sky. For example, it is a remarkable fact that the full Moon, which is half a degree across, can be covered by the width of a pencil held at arm's length. (Try it!) An outstretched palm will cover the main stars of Orion, and you can cover most of the stars in the Plough with your hand. An upright hand at arm's length has a height of about 14° from wrist to fingertips.

How can we judge whether a particular constellation covers a large or a small area of sky? The width of a fist held at arm's length makes a convenient distance indicator, about 7° wide. As a guide to scale, each of the constellation maps in this book carries an outline of a fist for comparison. If your hand is unusually big or small the comparison will not be exact, but it will still be a useful guide to the size of constellation you are looking for.

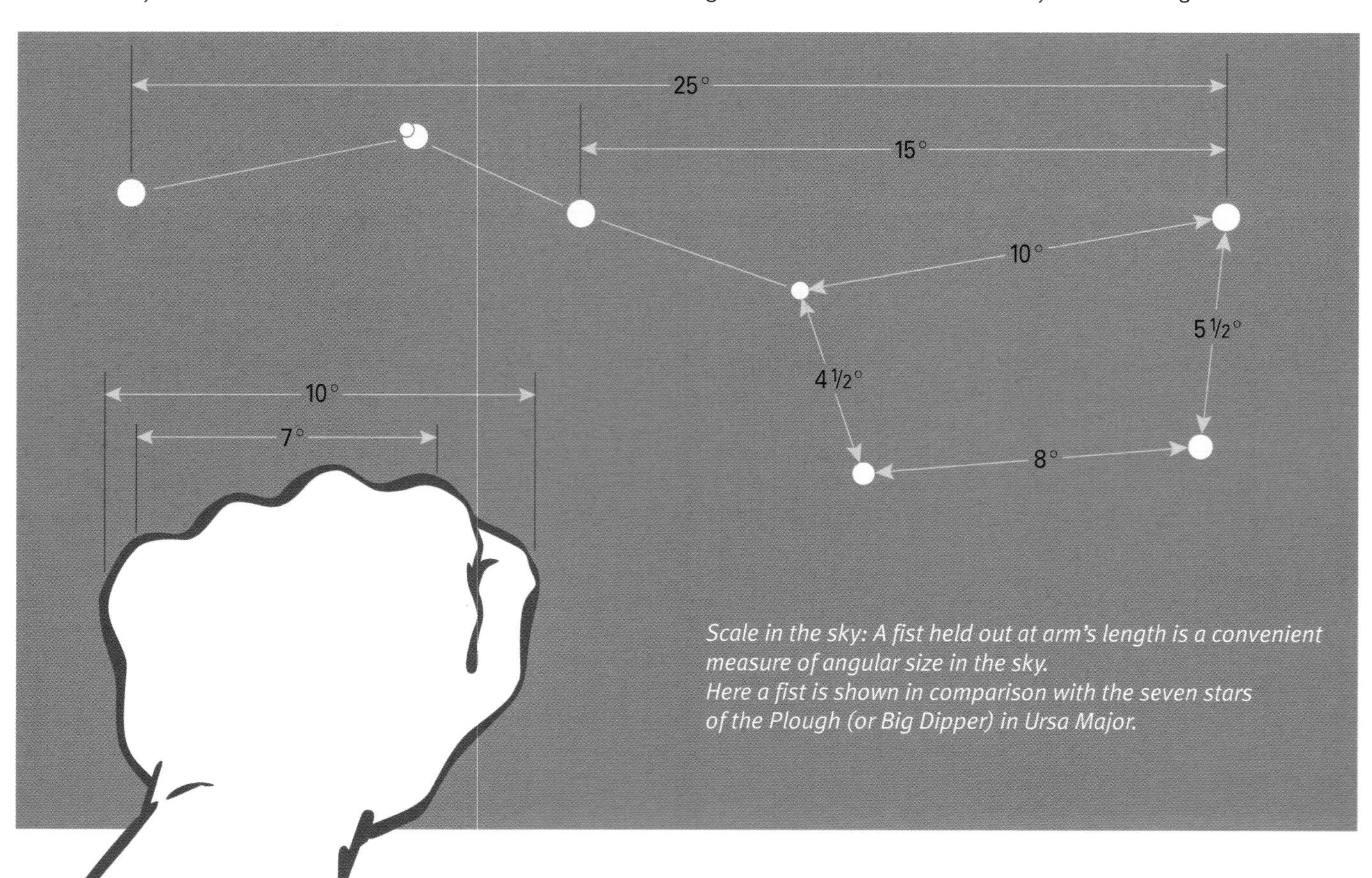

Scale in the sky: A fist held out at arm's length is a convenient measure of angular size in the sky.
Here a fist is shown in comparison with the seven stars of the Plough (or Big Dipper) in Ursa Major.

How the sky changes with the seasons

The sky is like a clock and a calendar. It changes in appearance with the time of night and with the seasons of the year. For example, look at the diagrams below of the area around the north pole of the sky. At 10 p.m. at the start of the year, the familiar saucepan-shape of seven stars known variously as the Plough or Big Dipper is standing on its handle to the right of Polaris, the north pole star. At the same time of night three months later, the Plough is almost directly overhead, while the W-shaped constellation of Cassiopeia sits low on the northern horizon. By July, the Plough appears to the left of Polaris. At 10 p.m. in October the Plough is scraping the northern horizon, while Cassiopeia rides high. In another three months the stars are back to where they started and the yearly cycle begins anew.

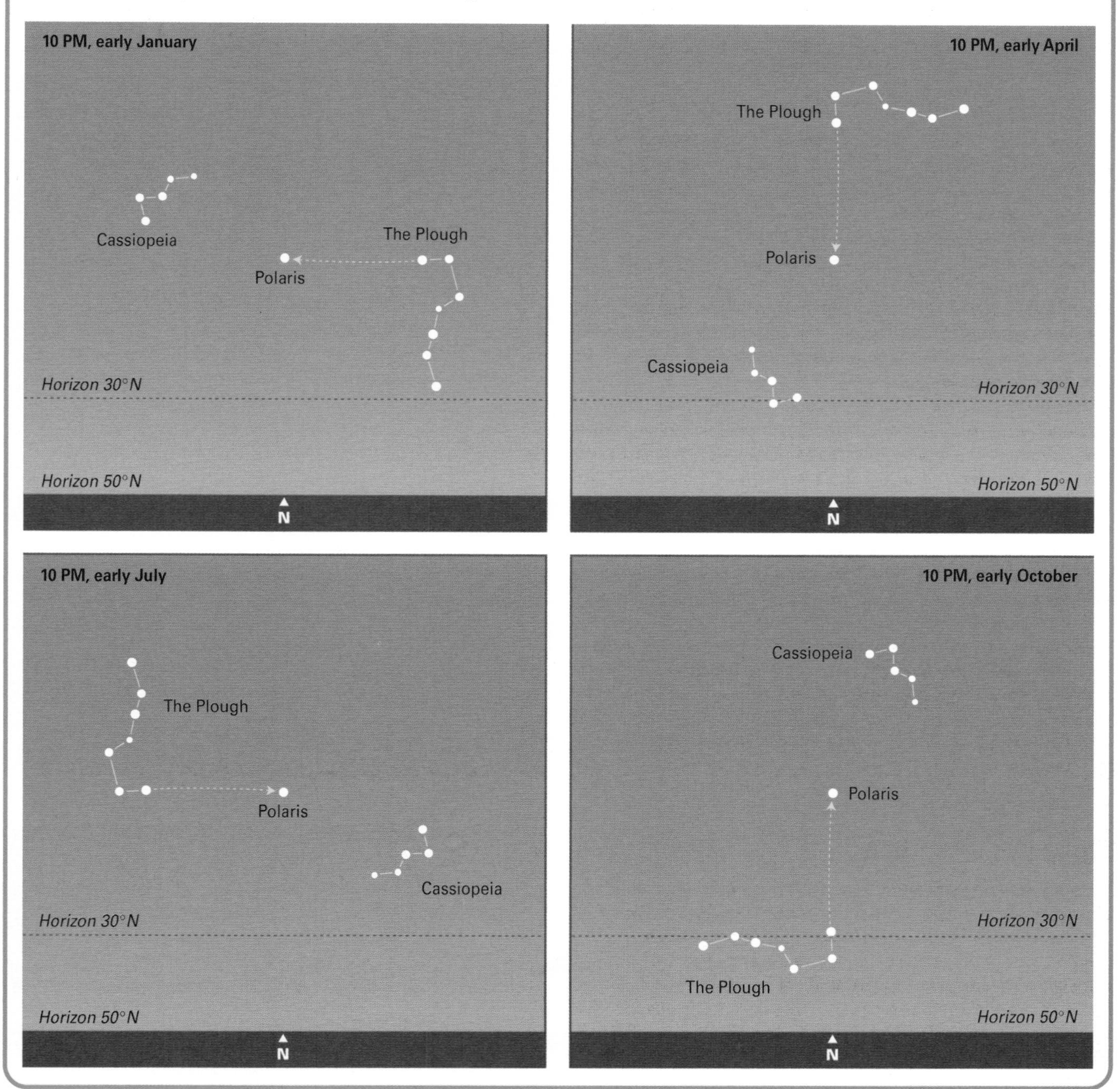

Greek alphabet

α	alpha	η	eta	ν	nu	τ	tau
β	beta	θ or ϑ	theta	ξ	xi	υ	upsilon
γ	gamma	ι	iota	ο	omicron	ϕ or φ	phi
δ	delta	κ	kappa	π	pi	χ	chi
ε	epsilon	λ	lambda	ρ	rho	ψ	psi
ζ	zeta	μ	mu	σ	sigma	ω	omega

Finding your way

To find your way in unfamiliar territory, you need a map and signposts. This book provides the maps. The signposts are in the sky, once you know where to look. Start with an easily recognizable pattern, such as the Plough in spring or Orion in winter, and work your way outwards from it to locate other constellations and bright stars, a technique known as star-hopping. While star-hopping around the sky you will find that there are many natural 'pointers' that direct you from constellation to constellation. Also, bright stars often form distinctive patterns of lines, triangles and squares that you can remember. The four small charts on these pages demonstrate some of the best ways of locating prominent stars and constellations in each season. As you navigate your way among the stars you will discover many more signposts of your own.

Signposts of spring

Start with the familiar saucepan-shape of the **Plough** or **Big Dipper**, which rides high in the sky on spring evenings. The seven stars of the saucepan are actually the most prominent members of the constellation Ursa Major, the Great Bear. From the diagram you can see that two stars of the saucepan's bowl – the ones that lie farthest from the handle – point towards the north pole star, **Polaris**. These two stars in the Plough are popularly known as **The Pointers**. If you extend the distance between them by about five times you will reach the pole star. Opposite Polaris from the Plough lie five stars that form a distinctive W-shape, which is the constellation of **Cassiopeia**.

Polaris is a star of moderate brightness lying in a somewhat blank region. It is not exactly at the north pole of the sky, but is situated about one degree (two Moon diameters) from it. During the night the stars circle around the north celestial pole (and hence around Polaris) as the Earth spins on its axis.

Go back to the Plough. If you extend the Pointers in the opposite direction, away from Polaris, you will come to the constellation of **Leo**, the Lion. This constellation is notable for the sickle-shape of stars, like a reversed question mark, that makes up the lion's head.

Now look at the handle of the Plough. Follow the curve made by the stars of the handle until you come to **Arcturus**, one of the brightest stars in the sky. Continue the curve and you reach the sparkling star **Spica**, in the constellation of Virgo. Note that Arcturus and Spica form a prominent triangle with **Regulus**, the brightest star in Leo.

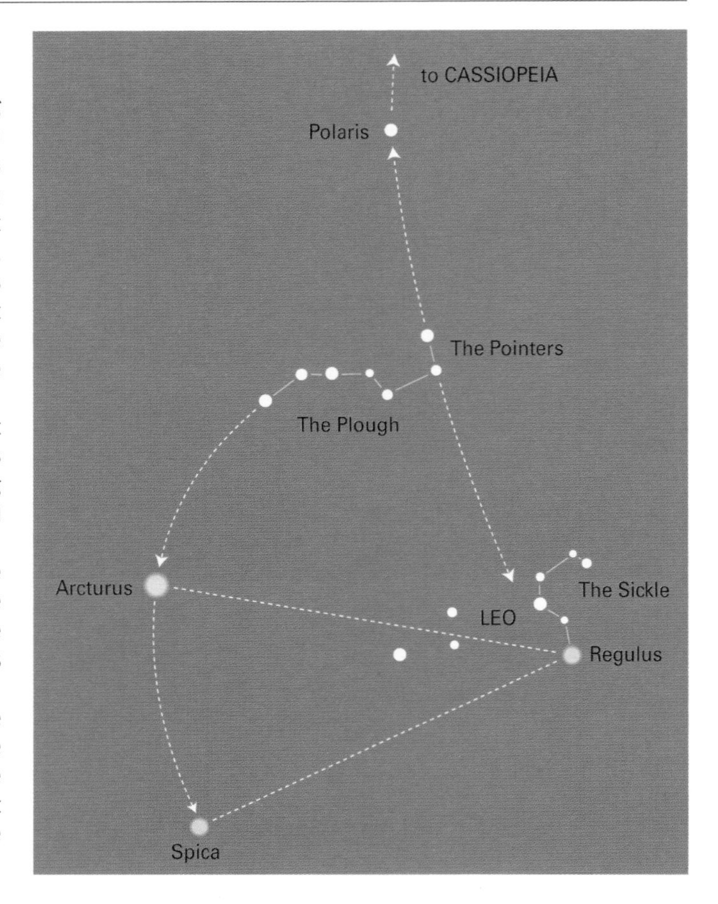

Signposts of summer

High in the sky on summer evenings lies an isosceles-shaped trio of bright stars known as the **Summer Triangle**. In order of decreasing brightness they are **Vega**, **Altair** and **Deneb**. Vega is the first star to appear as the sky darkens in July and August, shining overhead like a blue-white diamond.

Deneb is the brightest star in **Cygnus**, a constellation that represents a swan but which is better visualized as a cross, as shown on the diagram here. Deneb marks the head of the cross. The foot of the cross bisects a line between Vega and Altair, and points towards the bright star **Antares**, which appears low on the horizon from mid-northern latitudes. Antares lies due south as the sky darkens in July. It has a prominent red colour, and represents the heart of the Scorpion, Scorpius.

A line from Vega through Deneb directs you towards the **Square of Pegasus**, a quadrangle of four stars that is the centrepiece of the autumn sky. Apart from the Summer Triangle, the summer sky is remarkably bereft of prominent star patterns. Incidentally, despite its name the Summer Triangle remains visible well into the autumn.

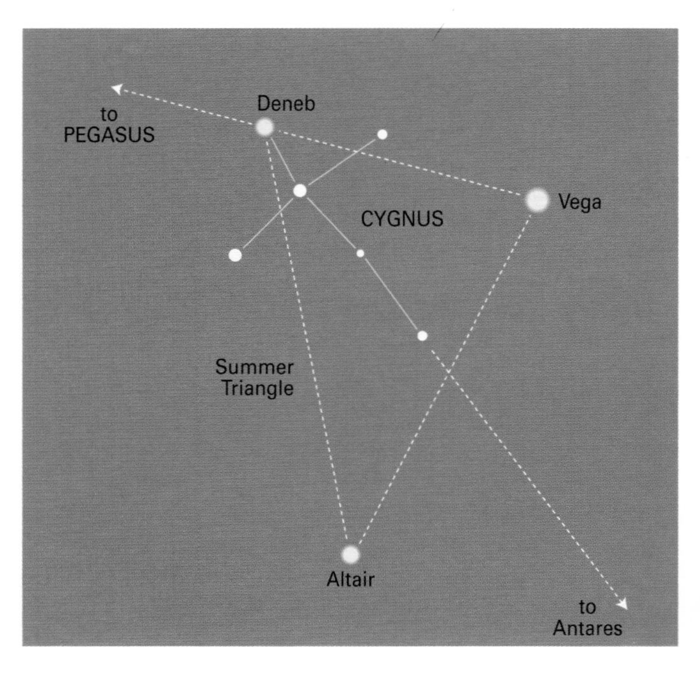

Signposts of autumn

As the Summer Triangle sets in the west on autumn evenings, the **Square of Pegasus** takes centre stage. The Square lies high in the south at 10 p.m. in mid-October, 8 p.m. in mid-November, and 6 p.m. in mid-December. The four stars that mark the corners of the Square are of only moderate brightness. They enclose a large area of sky that is almost entirely devoid of naked-eye stars.

The Square of Pegasus is like a keystone in the autumn sky, from which many surrounding stars and constellations can be identified. To the top right of the Square lie departing **Deneb** and the rest of the Summer Triangle. Between the Square and the pole star lies the W-shaped constellation of **Cassiopeia**. As the diagram at the right demonstrates, a line extended upwards from the left-hand side of the Square passes through the end of Cassiopeia's W and on to **Polaris**. Alternatively, you can use this line in reverse, running it from Polaris through Cassiopeia, to find the Great Square.

A line extended downwards from the right-hand side of the Square directs you to **Fomalhaut**, a bright star in the southern constellation of Piscis Austrinus, but often difficult to find from mid-northern latitudes because it is so close to the horizon.

Now look at Cassiopeia, riding high above you. A line drawn across the top of the W points to **Capella**, one of the prominent stars of the winter sky.

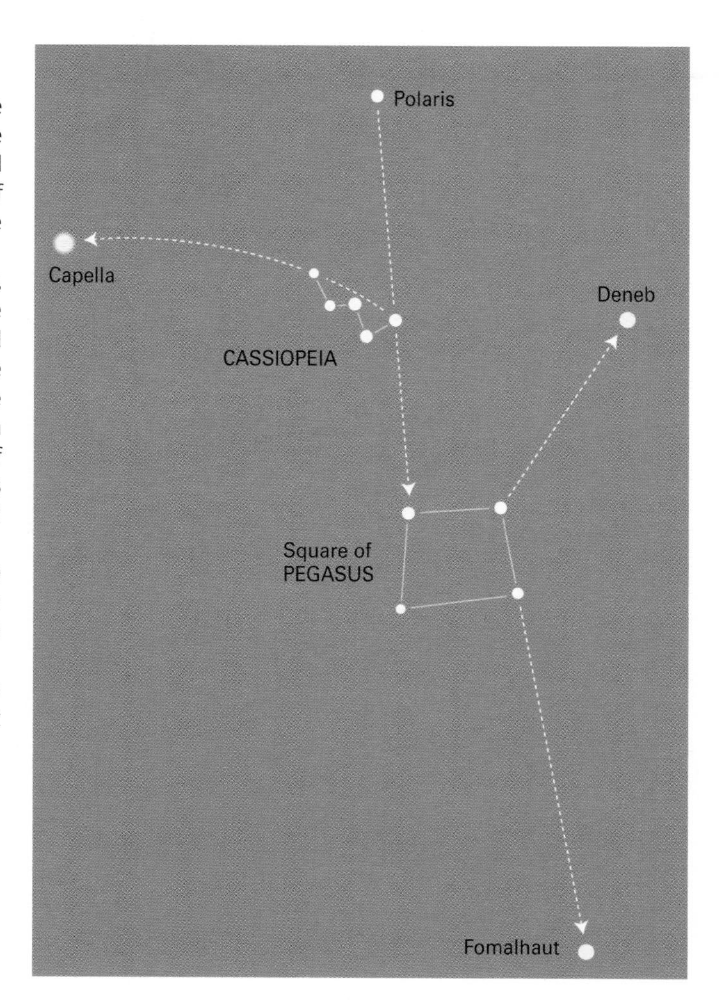

Signposts of winter

The sky in winter is more richly stocked with bright stars than in any other season of the year. The jewel in winter's crown is the brightest star in the entire sky, **Sirius**, which glitters due south at midnight at the beginning of January, at 10 p.m. at the beginning of February and at 8 p.m. at the beginning of March. Sirius lies at the southern apex of the **Winter Triangle** of brilliant stars, which is completed by **Procyon** to its upper left and **Betelgeuse** to its upper right.

Betelgeuse marks the top left of the constellation of **Orion**, a rectangular-shaped figure that is the easiest of the winter constellations to recognize. At the bottom right of Orion's rectangle is **Rigel**, a star slightly brighter than Betelgeuse. Two fainter stars complete the rectangle. Across the centre of Orion runs a distinctive line of three stars comprising **Orion's belt**.

To the top right of Orion is another prominent star, **Aldebaran**, which represents the glinting eye of Taurus, the Bull. Aldebaran is of similar brightness to Betelgeuse, and both stars have a noticeably orange tinge. Continue the line from Orion through Aldebaran and you will come to a hazy-looking knot of stars called the **Pleiades**, a star cluster that shows up well in binoculars.

Above Orion, almost directly between it and the north celestial pole, lies the bright star **Capella**. To the top left of Orion, above the Winter Triangle, lies a famous pair of stars called **Castor** and **Pollux**, the celestial twins in the constellation Gemini. Off to the left of Castor, Pollux and Procyon lies **Leo**, the Lion, which introduces us once again to the skies of spring.

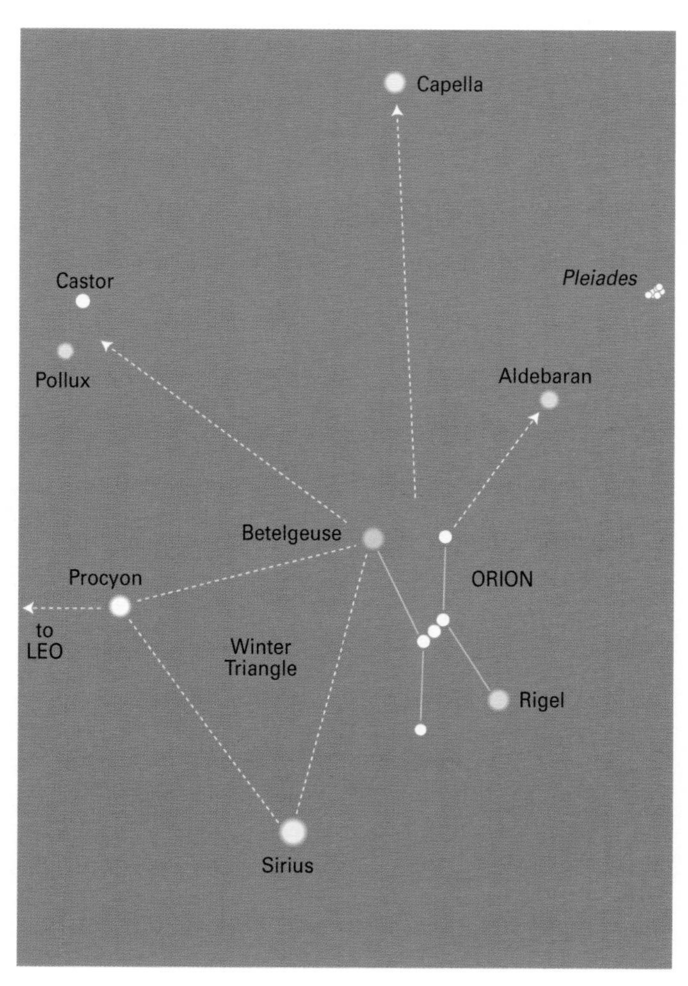

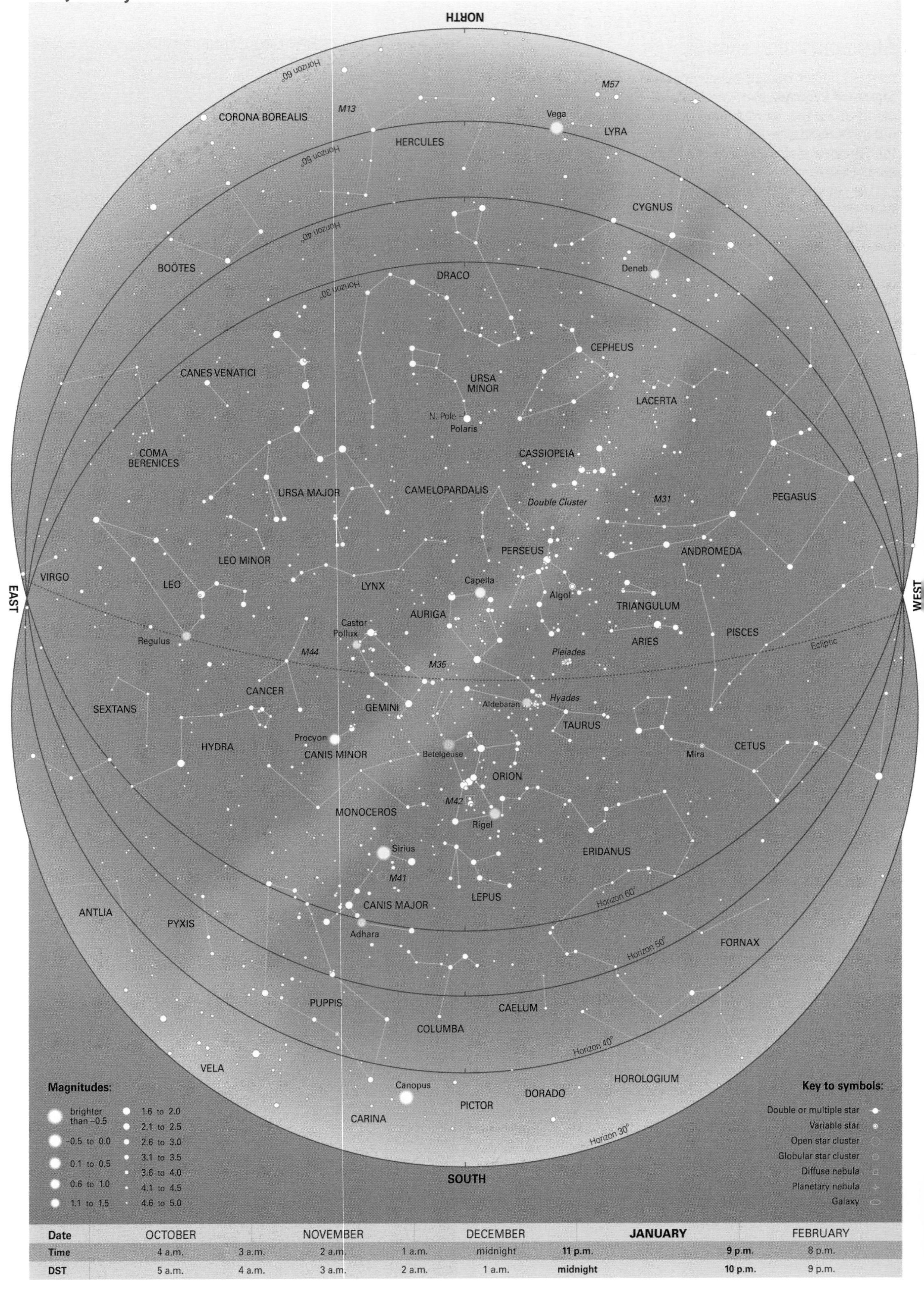

Date	OCTOBER		NOVEMBER		DECEMBER		**JANUARY**	FEBRUARY
Time	4 a.m.	3 a.m.	2 a.m.	1 a.m.	midnight	**11 p.m.**	**9 p.m.**	8 p.m.
DST	5 a.m.	4 a.m.	3 a.m.	2 a.m.	1 a.m.	**midnight**	**10 p.m.**	9 p.m.

January

Key Stars

Sirius, the jewel of the winter sky, glitters brilliantly in the south this month. It lies low in the sky for observers in high northern latitudes, and often appears to flash a multitude of colours as its light is refracted by the atmosphere. Sirius forms one vertex of the Winter Triangle of three bright stars, completed by red Betelgeuse to its upper right and white Procyon above and to its left. Capella appears directly overhead this month at 10 p.m. from mid-northern latitudes. Observers below latitude 37° north will find Canopus due south at 10 p.m., beneath Sirius (Canopus is below the horizon from more northerly latitudes).

The planets this month

Venus

2007 Emerges from evening twilight during the month, low in the western sky at mag. −3.9.

2008 In the south-east dawn sky at mag. −4.0. Ends the month close to fainter Jupiter.

2009 Brilliant in the south-west evening sky at mag. −4.5. Greatest elongation (maximum separation) from the Sun of 47° is on January 14.

2010 At superior conjunction (behind the Sun) on January 10 and invisible throughout the month.

2011 Dominates the south-east dawn sky at mag. −4.4. At greatest elongation (maximum separation) from the Sun of 47° on January 8.

Mars

2007 Emerges into the dawn sky during the month, low in the south-east at mag. 1.4.

2008 Visible all night in Taurus, fading rapidly from mag. −1.5 to −0.7.

2009 Too close to the Sun to be seen.

2010 Visible all night in Cancer, brightening to mag. −1.2 by month's end. At opposition (due south at midnight) on January 29, 99 million km (62 million miles) from Earth.

2011 Too close to the Sun to be seen.

Jupiter

2007 In the south-east dawn sky, in southern Ophiuchus at mag. −1.8.

2008 Emerges into the south-east dawn sky by month's end at mag. −1.9, below left of brilliant Venus.

2009 Rapidly swamped in evening twilight as it moves towards conjunction (behind the Sun) on January 24.

2010 In the south-west evening twilight at mag. −2.1, becoming very low by month's end.

2011 In the western evening sky in Pisces at mag. −2.2.

Saturn

2007 Visible for most of the night in Leo at mag. 0.1.

2008 Visible for most of the night under the body of Leo at mag. 0.5.

2009 In the eastern morning sky, stationary under the hind quarters of Leo at mag. 0.9.

2010 A morning object, stationary in Virgo at mag. 0.8.

2011 A morning object, stationary in Virgo at mag. 0.7.

Eclipses

Sun

2009 January 26. Annular solar eclipse, visible from the Indian Ocean, Sumatra and Borneo; maximum duration 7m 54s. A partial eclipse is visible from southern Africa, southern India, south-east Asia, western Indonesia, western and southern Australia.
http://sunearth.gsfc.nasa.gov/eclipse/solar.html

2010 January 15. Annular solar eclipse, visible from central and east Africa, southern India, Myanmar and China; maximum duration 11m 8s. A partial eclipse is visible from Africa, the Middle East, central and southeast Asia.
http://sunearth.gsfc.nasa.gov/eclipse/solar.html

2011 January 4. Partial solar eclipse, maximum 85%, visible from Europe, northern Africa, middle east and western Asia.
http://sunearth.gsfc.nasa.gov/eclipse/SEcat/SEdecade2011.html

January meteors

The year's most abundant meteor shower, the **Quadrantids**, is visible on January 3 and 4. Its peak of activity is much sharper than that of most showers, lasting only a few hours; the exact date and time of maximum varies from year to year. At best over 100 Quadrantids an hour can be seen, although the meteors of this shower are not as bright as other great displays such as the Perseids and Geminids. The Quadrantids radiate from the northern part of Boötes, near the handle of the Plough. This area of sky was once occupied by the now-defunct constellation Quadrans Muralis, the Mural Quadrant, from which the shower takes its name. Unfortunately this area of sky is not well placed for observation until after midnight, so Quadrantid watchers must be prepared to brave the small hours until dawn to see the shower at its best. Their parent body is thought to be an asteroid called 2003 EH_1, which was probably once part of a comet's nucleus.

Orion

Orion is the most glorious of all constellations. It glitters with bright stars and is replete with objects of interest for observers with all forms of optical equipment, including the humblest binoculars. It is an ideal place to begin a yearly tour of the heavens.

Orion is popularly personified as a hunter, giant or warrior. On old star charts he is depicted brandishing his club and shield against the snorting charge of neighbouring Taurus, the Bull. In Greek mythology Orion was the son of Poseidon. He was stung to death by a scorpion in retribution for his boastfulness. At the request of his lover, Artemis, he was placed in the sky so that he sets in the west as his slayer, in the form of the constellation Scorpius, rises in the east. In another story, Orion was smitten by the beauty of the seven nymphs known as the Pleiades, whom he began to pursue. To save the nymphs, Artemis turned them into a cluster of stars. The Pleiades star cluster lies in the neighbouring constellation of Taurus, and Orion still seems to chase them in his nightly course across the sky.

Orion is one of the easiest constellations to recognize, lying due south at 10 p.m. this month. Look first of all at Alpha (α) Orionis, better known as **Betelgeuse**, at the top left of the constellation. Note that it has a distinctly orange colour, which is brought out more strongly through binoculars. Betelgeuse marks the right shoulder of Orion. Its name is often mistranslated as 'armpit of the central one', but in fact its original Arabic title meant 'hand of Orion'.

Betelgeuse is the type of star known as a red supergiant. It is huge – over 500 times the diameter of the Sun, enough to fill the orbit of Jupiter – and fluctuates in size, leading to unpredictable changes in its brightness between about magnitude 0 and 1.3, although normally it hovers around magnitude 0.5. Keep an eye on the brightness changes of Betelgeuse by comparing it with other stars, particularly Aldebaran and Capella, from time to time. A measurement by the Hipparcos satellite places Betelgeuse 430 light years away, closer than the other bright stars in Orion.

Compare Betelgeuse with the blue-white star **Rigel** at the bottom right of the constellation. Rigel marks Orion's left leg and, not surprisingly, its name comes from the Arabic meaning just that – 'left leg'. Rigel, also known as Beta (β) Orionis, is the brightest star in Orion, magnitude 0.2. It, too, is a supergiant, but it has a much hotter surface, which accounts for its difference in colour from Betelgeuse. The temperature of Rigel's surface is 12,000 K (kelvin) while that of Betelgeuse is a relatively cool 3600 K. Rigel is about 770 light years away, nearly twice as far as Betelgeuse.

Rigel is worth looking at carefully for another reason: it is a double star. It has a companion star of magnitude 6.8 that is difficult to see in the smallest telescopes because of the glare from Rigel itself. The aperture needed to pick out this companion will depend on the steadiness of the atmosphere and how high Rigel is above the horizon. Probably at least 75 mm (3 inches) aperture is required, but the only way to be sure is to go out and see for yourself.

One distinctive feature of Orion is the line of three bright stars like shiny studs along his belt: Delta (δ) Orionis (Mintaka), Epsilon (ε) Orionis (Alnilam) and Zeta (ζ) Orionis (Alnitak). Each of these stars is of 2nd magnitude and they lie between about 800 and 1400 light years away. **Delta Orionis** has a wide 7th-magnitude companion star that is easily seen in small telescopes. At the left end of the belt, **Zeta Orionis** is a double star that presents a challenge for amateur telescopes. A 4th-magnitude companion lies 2.3 seconds of arc from it; normally, stars this far apart should just be divisible in an aperture of 50 mm (2 inches), but the difference in brightness between them means that in practice at least 75 mm (3 inches) aperture is needed to separate them, as well as a night of steady seeing.

A strip of faint nebulosity, known as **IC 434**, extends southwards from Zeta Orionis. On one side of this is the celebrated **Horsehead Nebula**, formed by a cloud of dark dust that is silhouetted against the glowing hydrogen gas behind it. Unfortunately, the Horsehead is virtually impossible to see even with large telescopes, but it shows up well on long-exposure photographs, looking like a celestial chess piece.

A survey of Orion's stellar treasures would be incomplete without a visit to **Sigma** (σ) **Orionis**, lying just below the left-most star of the belt. To the naked eye Sigma Orionis appears as an unremarkable star of 4th magnitude, but small telescopes reveal it to be flanked by three fainter stars, looking like a planet with moons. Also visible in the same field of view is a triple star, **Struve 761**, which consists of a narrow triangle of 8th- and 9th-magnitude stars. This rich telescopic sight is one of Orion's unexpected delights.

One reason for Orion's prominence is that most of its stars lie in a region of our local spiral arm of the Galaxy where new stars are still being born. The centre of starbirth in this region is the **Orion Nebula**, 1500 light years away, which appears as a misty patch to the naked eye; it marks the sword of Orion, hanging from his belt (*see panel below*).

The Orion Nebula is about 20 light years in diameter and contains enough gas to make thousands of stars. If we could turn back the clock 5000 million years to the birth of the Sun we would find that our region of space looked much like the Orion Nebula.

The Orion Nebula

The Orion Nebula is a luminous cloud of gas, the finest object of its kind in the heavens, also known as M42 and NGC 1976. Wreaths of ghostly glowing gas spread over a degree of sky, twice the apparent diameter of the full Moon, making a sight not to be missed in any instrument. To its north is a smaller patch of gas, **M43**, alias NGC 1982, actually part of the same cloud. Because the Nebula is so large, binoculars are excellent for viewing it. Embedded at the Nebula's centre is a 5th-magnitude star, **Theta-1** (θ^1) **Orionis**. This star was born recently from the surrounding gas, and it illuminates the Nebula. Home in on Theta-1 Orionis with a small telescope and you will see why it is popularly known as the **Trapezium**: it consists of four stars of magnitudes 5.1, 6.7, 6.7 and 8.0, arranged in a rectangle. The Trapezium is almost at the tip of a dark wedge in the Nebula known as the **Fish Mouth**. To the lower left of the Trapezium, binoculars show **Theta-2** (θ^2) **Orionis**, a wide double star of magnitudes 5.0 and 6.4.

This whole area is a complex of nebulosity and young, hot stars, which together comprise the sword of Orion. At the tip of the sword, on the southern edge of the Orion Nebula, is **Iota** (ι) **Orionis**, the brightest star in the sword. Iota Orionis is a hot, blue-white giant of 3rd magnitude with a 7th-magnitude companion shown by small telescopes. To the lower right of Iota lies **Struve 747**, an easy pair of 5th- and 6th-magnitude white stars for small telescopes. Above the Orion Nebula lies a smaller smudge of nebulosity, **NGC 1977**, containing the 5th-magnitude star 42 Orionis. (Its apparent neighbour, 45 Orionis, is an unrelated foreground object.) Farther north, binoculars reveal **NGC 1981**, a scattered handful of stars of 6th magnitude and fainter, marking the top of the sword handle.

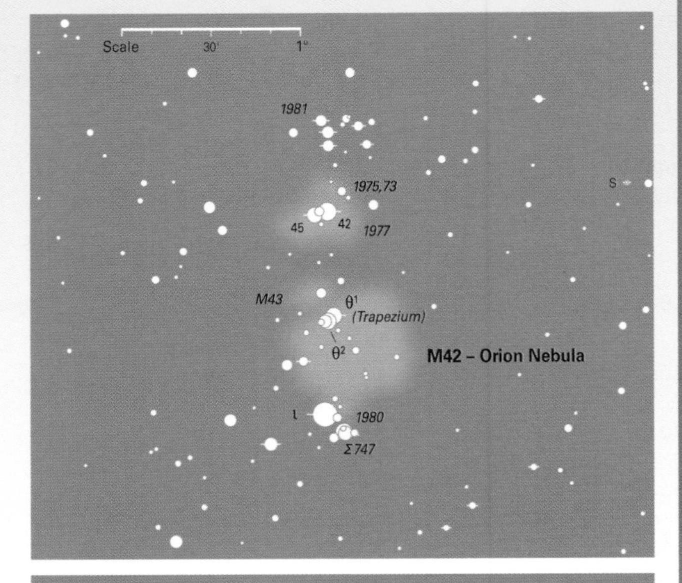

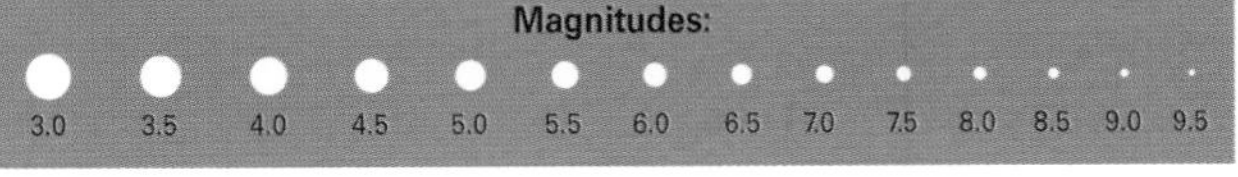

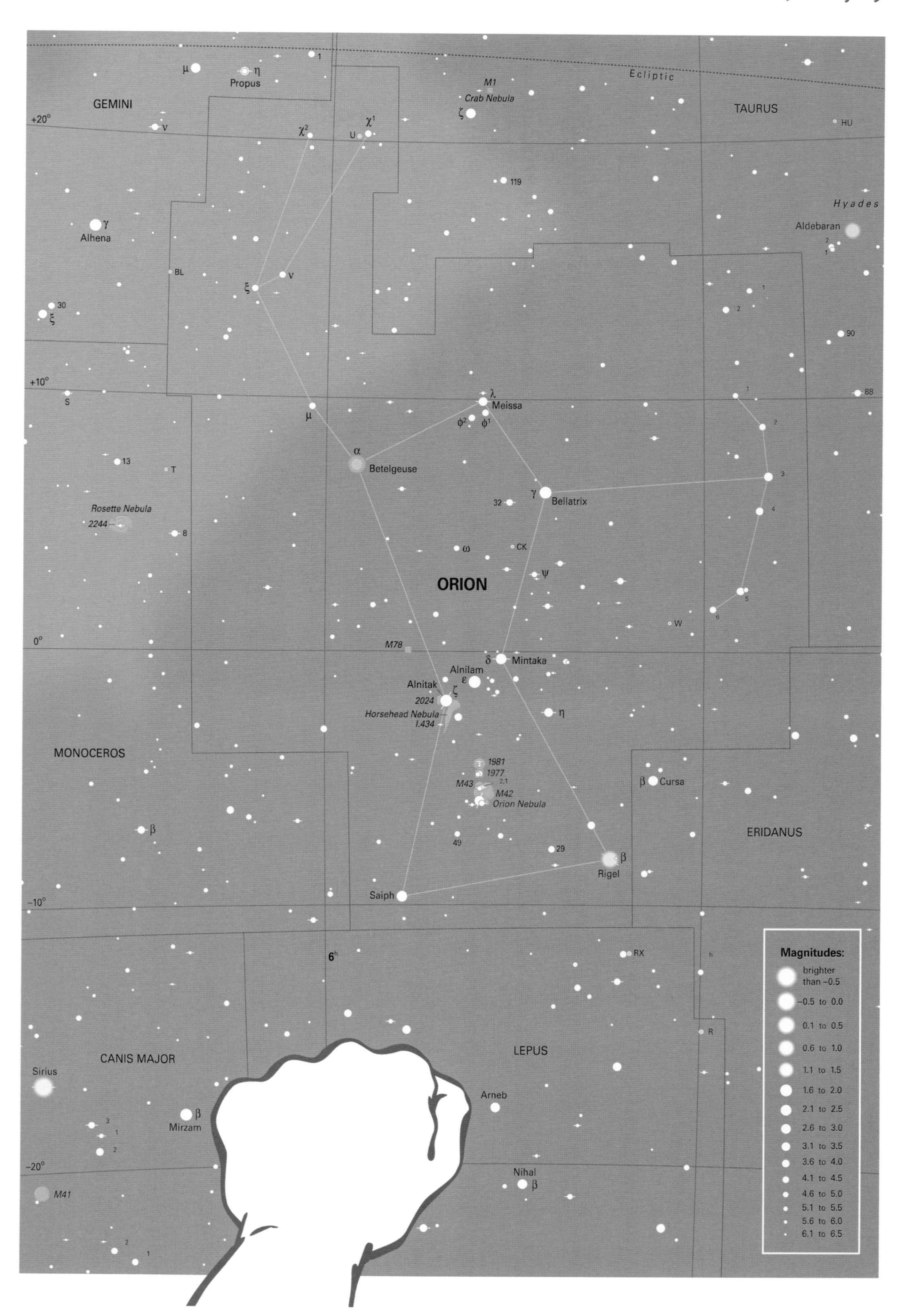
GEMINI
Propus
Alhena
Ecliptic
M1
Crab Nebula
TAURUS
Hyades
Aldebaran
Meissa
Betelgeuse
Bellatrix
Rosette Nebula
2244
ORION
M78
Mintaka
Alnilam
Alnitak
2024
Horsehead Nebula
I.434
1981
1977
M43
M42
Orion Nebula
Cursa
MONOCEROS
ERIDANUS
Rigel
Saiph
CANIS MAJOR
LEPUS
Sirius
Arneb
Mirzam
Nihal
M41
Magnitudes:
brighter than −0.5
−0.5 to 0.0
0.1 to 0.5
0.6 to 1.0
1.1 to 1.5
1.6 to 2.0
2.1 to 2.5
2.6 to 3.0
3.1 to 3.5
3.6 to 4.0
4.1 to 4.5
4.6 to 5.0
5.1 to 5.5
5.6 to 6.0
6.1 to 6.5

20 February

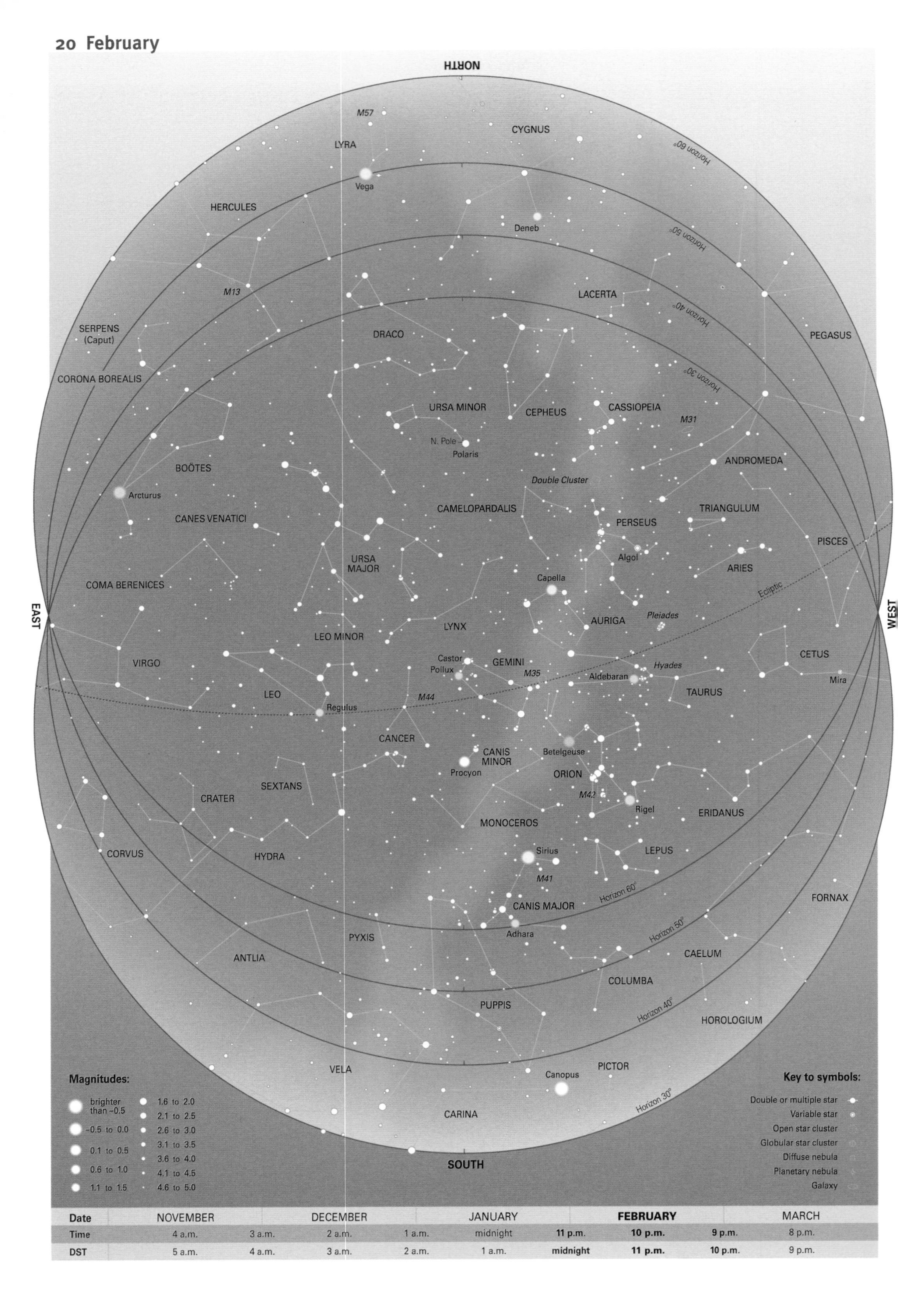

Date	NOVEMBER		DECEMBER		JANUARY		**FEBRUARY**		MARCH
Time	4 a.m.	3 a.m.	2 a.m.	1 a.m.	midnight	**11 p.m.**	**10 p.m.**	**9 p.m.**	8 p.m.
DST	5 a.m.	4 a.m.	3 a.m.	2 a.m.	1 a.m.	**midnight**	**11 p.m.**	**10 p.m.**	9 p.m.

February

Key Stars

The Winter Triangle formed by Sirius, Procyon and Betelgeuse remains prominent in the evening sky this month. Other stars well-placed for observation are Rigel, in Orion, and Aldebaran, in Taurus, both in the south-west. Castor and Pollux, the twin stars of Gemini, are high in the sky at 10 p.m. At the same time, orange Arcturus is rising on the eastern horizon.

The planets this month

Venus

2007 Low in the western evening sky at mag. –3.9.
2008 Low in the south-east dawn sky at mag. –3.9. Starts the month close to fainter Jupiter but is then lost in twilight.
2009 Prominent in the western evening twilight throughout the month at a dazzling mag. –4.6.
2010 Too close to the Sun for observation all month.
2011 In the south-east dawn sky at mag. –4.2.

Mars

2007 Low in the south-east dawn sky, moving from Sagittarius into Capricornus at mag. 1.3.
2008 Visible for most of the night in Taurus, fading from mag. –0.6 to +0.1.
2009 Submerged in dawn twilight throughout the month.
2010 Visible all night in Cancer, fading from mag. –1.2 to –0.7.
2011 At conjunction (behind the Sun) on February 4 and invisible all month.

Jupiter

2007 A morning object in southern Ophiuchus at mag. –2.0.
2008 In the south-east dawn sky at mag. –1.9, passing brilliant Venus at the start of the month.
2009 Emerges into the south-east dawn sky by month's end at mag. –1.9, but remains very low from northerly latitudes.
2010 Starts the month low in the west after sunset but rapidly becomes lost in evening twilight. At conjunction (behind the Sun) on February 28.
2011 In the western evening sky in Pisces at mag. –2.1.

Saturn

2007 Visible all night in Leo at mag. 0.0. At opposition (due south at midnight) on February 10, 1227 million km (763 million miles) from Earth.
2008 Visible all night under the body of Leo. At opposition (due south at midnight) on February 24 at mag. 0.2, 1240 million km (771 million miles) from Earth.
2009 Visible for most of the night under the hind quarters of Leo, brightening from mag. 0.8 to 0.5 during the month.
2010 Visible for most of the night in Virgo at mag. 0.7.
2011 A morning object, almost stationary in Virgo at mag. 0.5.

Eclipses

Sun

2008 February 7. Annular solar eclipse, visible from Antarctica and the South Pacific; maximum duration 2m 8s. A partial eclipse is visible from Antarctica, New Zealand and south-east Australia.
http://sunearth.gsfc.nasa.gov/eclipse/solar.html

Moon

2008 February 21. Total lunar eclipse, visible from western Asia, the Middle East, Europe, Africa, North and South America. Totality starts 03.00 GMT, ends 03.51 GMT.
http://sunearth.gsfc.nasa.gov/eclipse/lunar.html

Canis Major

In the south this month sits the constellation Canis Major, the Greater Dog, home of **Sirius**, the brightest star in the entire sky. Canis Major and nearby Canis Minor represent two dogs following at the heels of Orion. Surprisingly, the only object of significance in the Lesser Dog, Canis Minor, is its brightest star, Procyon; hence Canis Minor might be termed the lone-star constellation. By contrast, Canis Major is packed with bright stars. On old maps, the dog is depicted as standing on his hind legs, with flaming Sirius (popularly known as the Dog Star) marking his snout.

Sirius has a magnitude of –1.4. Its name comes from the Greek meaning searing or scorching, for in ancient times it was actually thought to be a source of heat. The sweltering 'dog days' of high summer were attributed to the Dog Star, for they occurred when Sirius lay close to the Sun. Hesiod, one of the earliest Greek poets, wrote of 'heads and limbs drained dry by Sirius'. Because of the outstanding brilliance of Sirius, its yearly passage around the sky was used as a calendar-marker from at least the time of the ancient Egyptians, over 2000 years BC. By chance, it now lies almost exactly due south at midnight on January 1.

Sirius shines twice as brightly in our skies as the second most prominent star, **Canopus**, mag. –0.6, which can be seen below it at this time of year if you are south of latitude 37° north. The exceptional brilliance of Sirius is due to a combination of its light output and its relative closeness to us. Sirius is roughly twice the mass and twice the diameter of the Sun, and gives out more than 20 times as much light. That by itself is not exceptional. What helps make Sirius so bright in our skies is that it is 8.6 light years away, the fifth-closest star to the Sun. Hence it outshines more powerful stars, such as Betelgeuse in Orion, which are much more distant. Of the stars visible to the naked eye, only Alpha Centauri in the southern hemisphere is closer to us than Sirius.

In reality Sirius is a blue-white star, but it twinkles a multitude of colours as its light is broken up by air currents in the Earth's atmosphere on frosty winter nights. In binoculars or a small telescope it is dazzling. Sirius has a bizarre 8th-magnitude companion, a small and dense star of the type known as a *white dwarf* (see box on next page). This tiny companion of the Dog Star has been nicknamed the Pup. Unfortunately, it is so close to Sirius that a large telescope is needed to show it.

Beta (β) Canis Majoris is a blue giant star 500 light years away, magnitude 2.0. It bears the Arabic name Mirzam, meaning 'the announcer', from the fact that its rising heralds the appearance of Sirius. The second-brightest star in the constellation, Epsilon (ε) Canis Majoris, magnitude 1.5, is another blue giant, 430 light years away.

The real superstar of the constellation, though, is easily overlooked: **Eta (η) Canis Majoris**, a brilliant supergiant 100,000 times more luminous than the Sun but appearing to us only a modest magnitude 2.4 because it is so far away, about 3000 light years. Almost as impressive is **Delta (δ) Canis Majoris**, which the Arabs named Wezen, meaning 'weight'. The legend behind this odd name has been lost in the mists of time; although the Arabs

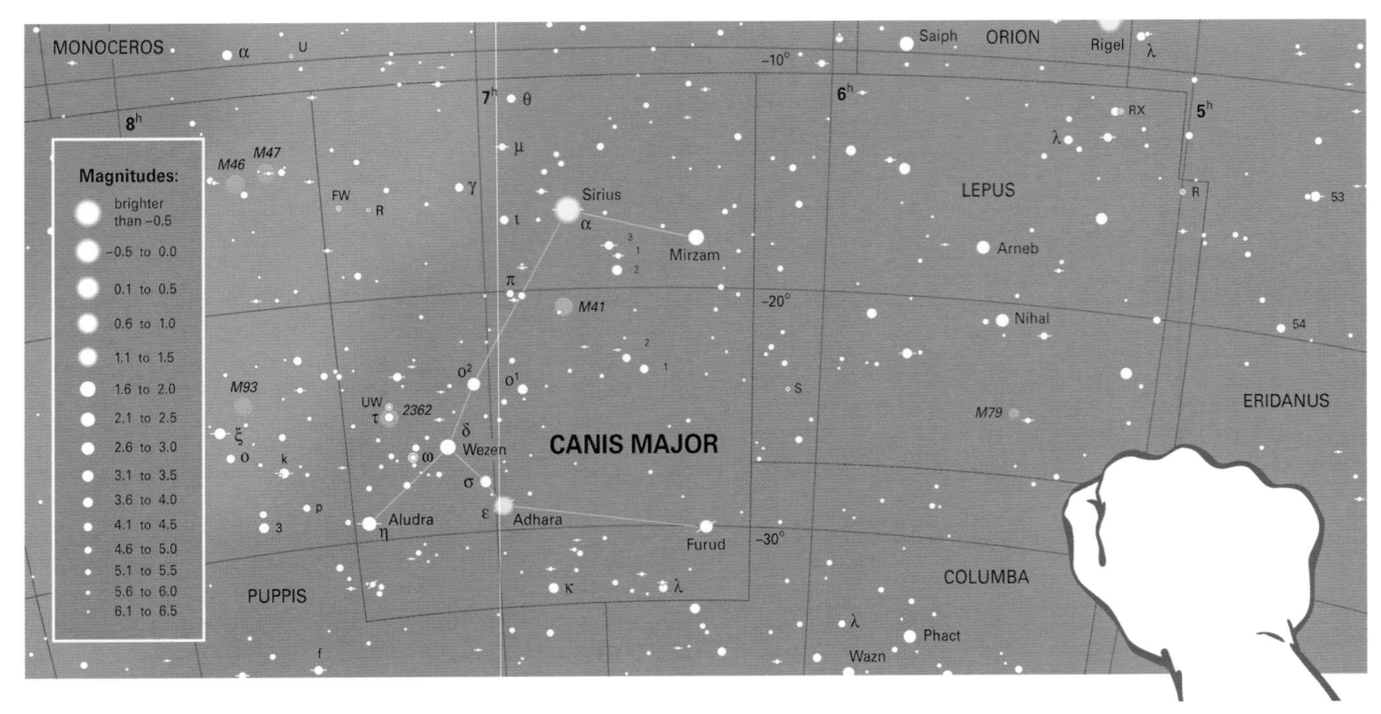

could not have known, the name is appropriate, since the star is indeed weighty, with an estimated mass at least 10 times that of the Sun, but it would have been even better suited to Eta Canis Majoris, which has turned out to be heavier still. Delta Canis Majoris is also a supergiant, in this case giving out about 50,000 times as much light as the Sun and appearing of magnitude 1.8 despite being nearly 2000 light years away. If Sirius were removed to the same distance as Delta or Eta Canis Majoris it would appear of 11th or 10th magnitude respectively, way below naked-eye visibility. Conversely, if these two stars were as close to us as Sirius they would each shine as brilliantly as a half Moon, so the night sky would never be truly dark when they were above the horizon.

Canis Major contains two attractive star clusters for small instruments. Just south of Sirius is **M41**, covering the same area of sky as the full Moon. Under good conditions it can even be seen as a fuzzy spot by the naked eye. M41 contains about 50 stars, the brightest of them a 7th-magnitude orange giant. Binoculars pick out the brightest members, and many more are visible in small telescopes. Low magnification is needed to fit all the cluster into the field of view. Note how the stars are arranged in chains, a common effect in clusters. Observers in high northern latitudes will find that the proximity of M41 to the horizon dims its splendour.

Much more compact than M41 is the star cluster **NGC 2362**, centred on the 4th-magnitude blue supergiant **Tau** (τ) **Canis Majoris**. NGC 2362 contains about 50 members, many of them visible in small telescopes. Tau Canis Majoris itself has a light output of 30,000 Suns. This star and the surrounding cluster are 5000 light years away.

White dwarfs

In 1862 the American astronomer Alvan G. Clark first saw the faint companion of Sirius while testing a new 470-mm (18½-inch) telescope. But this companion star, **Sirius B**, was a puzzle. Its surface was hotter than that of the Sun, yet the star itself was much dimmer than the Sun. That meant it must be very small – as we now know, it is only about 1 per cent of the Sun's diameter, similar in size to the Earth, which is astoundingly small by stellar standards. Such a tiny, hot star is termed a white dwarf.

Yet the puzzle did not end there. Sirius B was found to contain as much mass as the Sun. That much mass packed into such a small ball meant that Sirius B must be incredibly dense, far denser than any substance known on Earth. In fact, a spoonful of material from Sirius B would weigh many tonnes. By coincidence, the brightest star in Canis Minor, Procyon, is also accompanied by a white dwarf, fainter and even more difficult to see than the companion of Sirius. White dwarfs are now known to be the dying remnants of stars like the Sun. The companions of Sirius and Procyon have both seen better days.

Sirius B orbits Sirius A once every 50 years (*see diagram at right*). The two stars are always too close for Sirius B to be seen in small telescopes. Even when the two are at their widest separation, around AD 2025, a telescope of at least 200 mm (8 inches) aperture and steady air will be needed to pick out the faint spark of Sirius B from the glare of its brilliant neighbour.

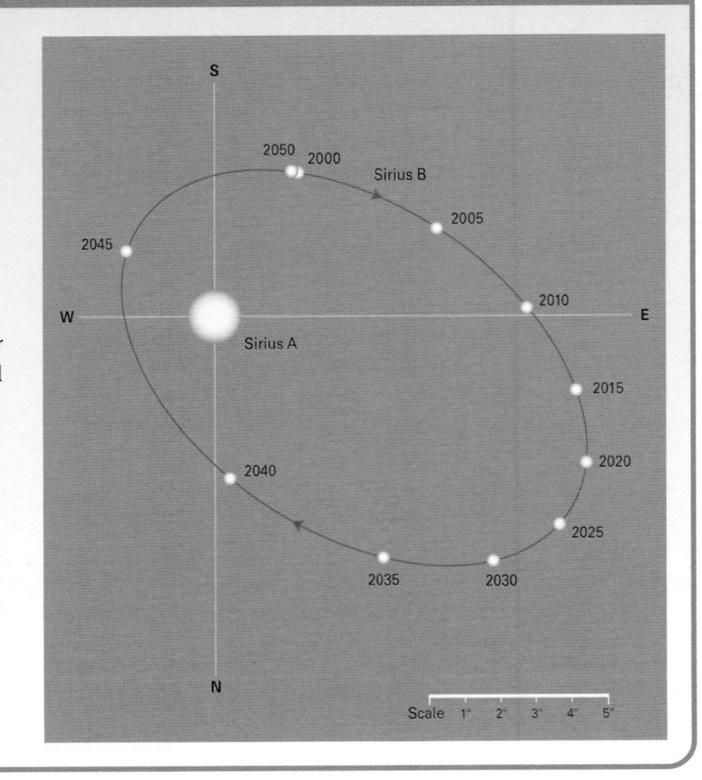

Monoceros

Between the stars of the Winter Triangle lies the faint but fascinating constellation of Monoceros, the Unicorn, introduced on star maps in the seventeenth century. Often overlooked in favour of its brilliant neighbours, Monoceros none the less contains some classic objects.

Among them is perhaps the finest triple star for small telescopes, **Beta (β) Monocerotis**. It appears double under low magnification, but higher power divides the fainter companion. The three stars, all of 5th magnitude, form an arc. **8 Monocerotis**, also known as Epsilon (ε) Monocerotis, is a double star easily separated by small telescopes, consisting of blue-white and yellow components of magnitudes 4.4 and 6.7, set among an attractive scattering of fainter stars. The two component stars are unrelated.

Monoceros contains two celebrated star clusters, both enveloped by faint nebulosity from which they have recently formed. **NGC 2244** is a widespread group of a dozen or so stars, 6th magnitude and fainter, covering an area similar to the apparent width of the Moon and best seen in binoculars or wide-field telescopes with low magnification. The hidden treasure of this object is the surrounding gas cloud, known as the **Rosette Nebula**, beautiful on long-exposure photographs. Unfortunately it is too faint to be seen through small telescopes, although under excellent conditions it might be glimpsed through large binoculars as a faint halo, more than two Moon diameters wide. It lies about 5000 light years away and is one of the youngest clusters known, with an age of about 3 million years.

About two degrees from NGC 2244 lies 6th-magnitude **Plaskett's Star**, named after the Canadian astronomer John Stanley Plaskett who found in 1922 that it is the heaviest pair of stars known. According to latest measurements, both stars are blue supergiants about 50 times as massive as the Sun, which places them firmly in the super-heavyweight class. The pair are so close together that they cannot be seen individually in any telescope. Only by analysing light from Plaskett's Star through a spectroscope can astronomers tell that two stars, not one, are present, and that they orbit each other every 14 days. Stars of such immense mass do not live for long, so Plaskett's Star must be very young, probably less than a million years old.

Another of the secret treasures of Monoceros is the star cluster **NGC 2264**. Small telescopes and binoculars show about a dozen stars arranged in the shape of an arrowhead, with numerous fainter members. Brightest of the cluster is **S Monocerotis**, an intensely hot blue-white star of magnitude 4.7 but slightly variable. NGC 2264 is enveloped in a glowing gas cloud, too faint to see in small telescopes but well known from long-exposure photographs. Into its southern end intrudes a cone-shaped wedge of dark dust which gives the object its popular name, the **Cone Nebula**. NGC 2264 and the Cone Nebula lie about 2500 light years away.

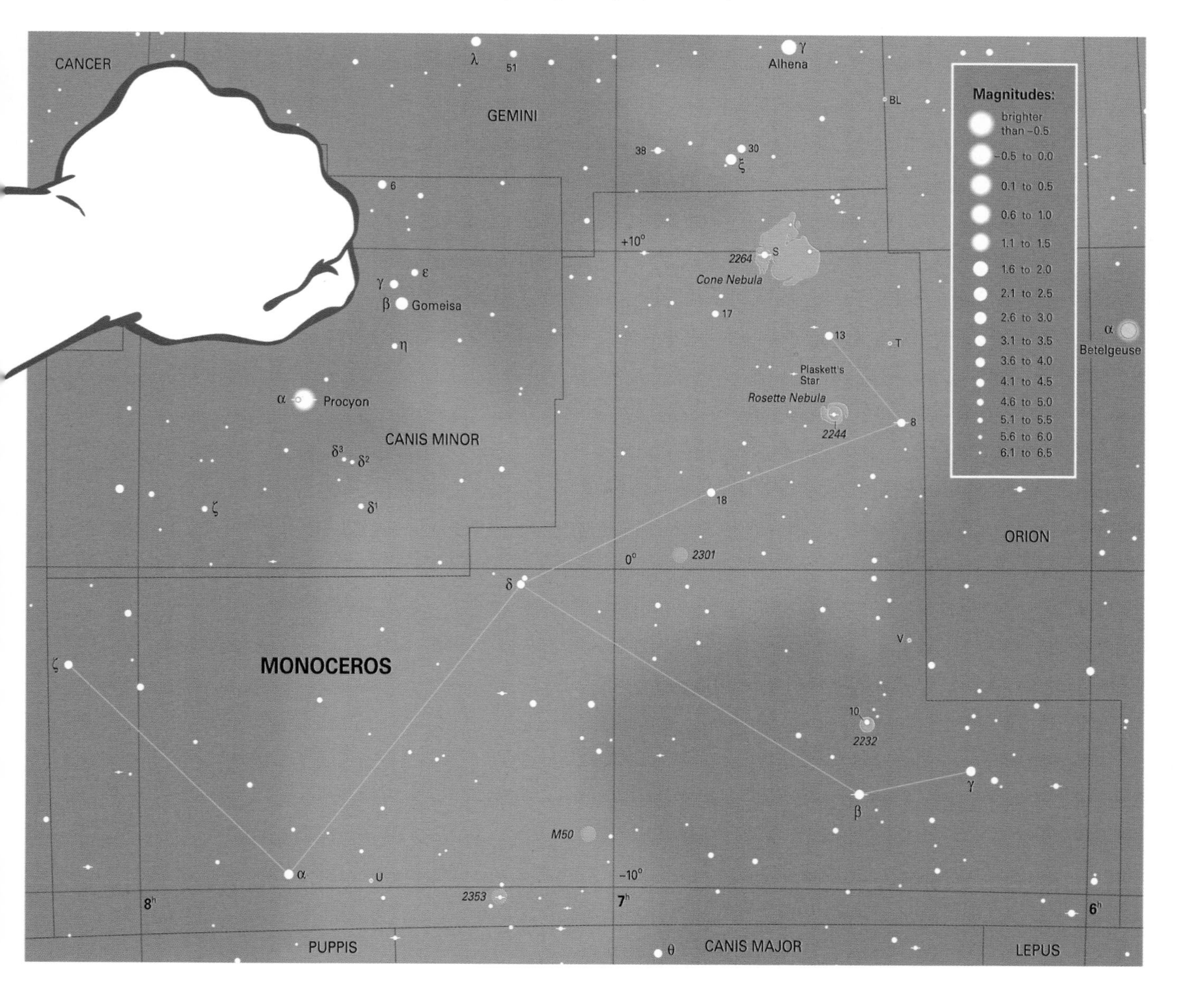

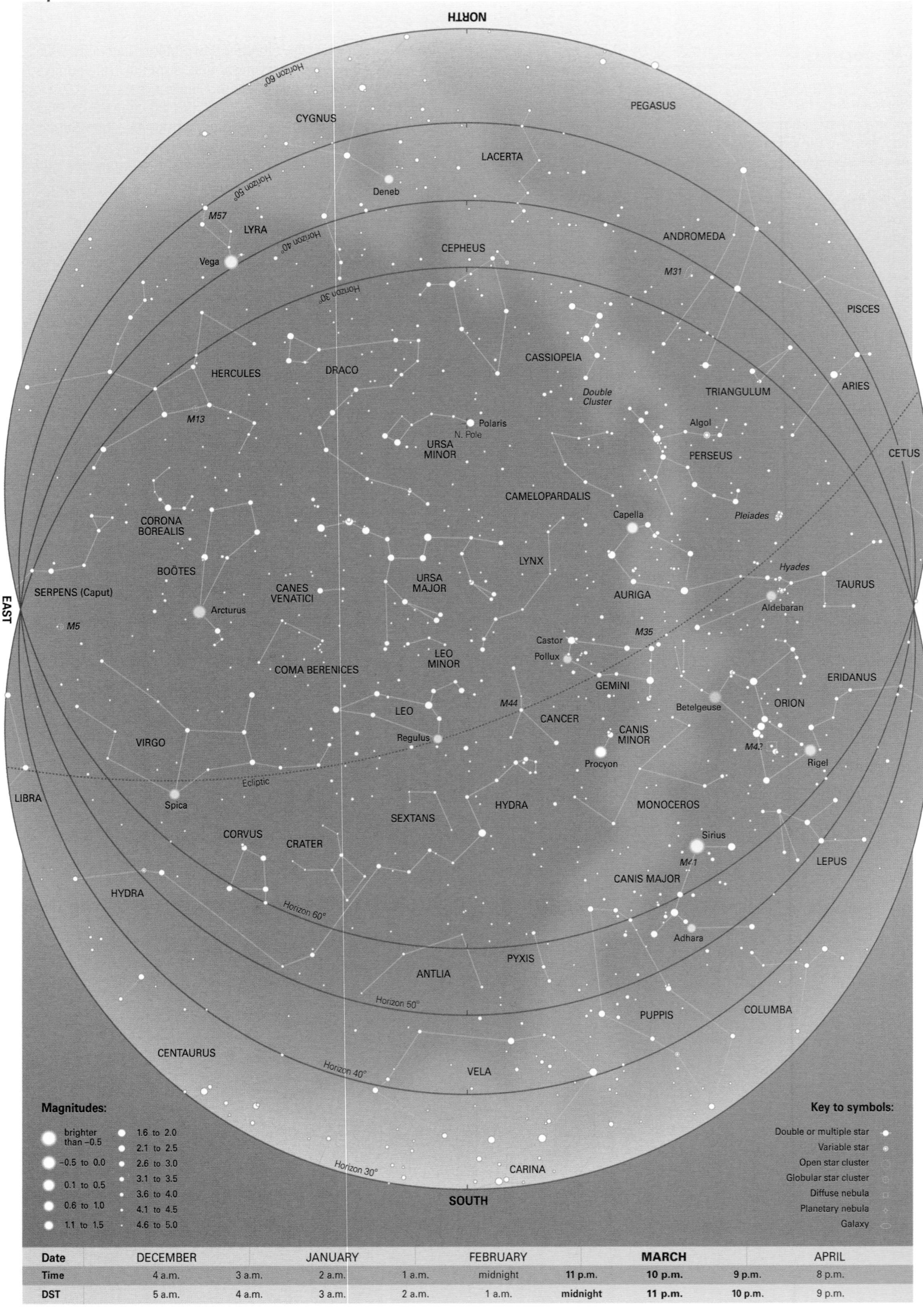

Date	DECEMBER		JANUARY		FEBRUARY		MARCH		APRIL
Time	4 a.m.	3 a.m.	2 a.m.	1 a.m.	midnight	**11 p.m.**	**10 p.m.**	**9 p.m.**	8 p.m.
DST	5 a.m.	4 a.m.	3 a.m.	2 a.m.	1 a.m.	**midnight**	**11 p.m.**	**10 p.m.**	9 p.m.

March

Key Stars

Taurus, Orion and the Winter Triangle are sinking towards the western horizon by 10 p.m. this month. Castor and Pollux are still well displayed, and Capella glints high in the north-west. Regulus, brightest star of Leo, shines high in the south around 10 p.m. Below it is the straggling figure of Hydra, a region bereft of bright stars. The seven stars that make up the familiar shape of the Plough, or Big Dipper, are well placed for viewing this month. On the eastern horizon, Spica and Arcturus are rising, heralding the approach of spring.

The planets this month

Venus

2007 Prominent in the western sky after sunset at mag. –4.0.
2008 Lost in dawn twilight throughout the month.
2009 Low in the western evening sky at mag. –4.5 for the first half of the month, and then swamped by twilight. Reaches inferior conjunction (between Earth and Sun) on March 27.
2010 Emerges into the evening twilight during the month, low in the west at mag. –3.9.
2011 Low in the south-east dawn sky at mag. –4.1.

Mars

2007 Low in the south-east dawn sky in Capricornus at mag. 1.2.
2008 Moves from Taurus into Gemini at mag. 0.5.
2009 Remains submerged in dawn twilight throughout the month.
2010 Visible for most of the night in Cancer, fading from mag. –0.6 to +0.1.
2011 Too close to the Sun to be seen.

Jupiter

2007 A morning object, stationary in southern Ophiuchus at mag. –2.2.
2008 A morning object in Sagittarius at mag. –2.0.
2009 Low in the south-east dawn sky at mag. –2.0.
2010 Too close to the Sun for observation.
2011 Moves rapidly towards the Sun, becoming lost in the western evening twilight around mid-month.

Saturn

2007 Visible all night between the stars of Leo and Cancer at mag. 0.1.
2008 Visible all night in Leo at mag 0.3.
2009 Visible all night under the body of Leo at mag 0.5. At opposition (due south at midnight) on March 8, 1256 million km (780 million miles) from Earth.
2010 Visible all night in Virgo. At opposition (due south at midnight) on March 22 at mag. 0.5, 1272 million km (791 million miles) from Earth.
2011 Visible for most of the night in Virgo at mag. 0.4.

Eclipses

Sun

2007 March 19. Partial solar eclipse, maximum 87%, visible from eastern Asia.
`http://sunearth.gsfc.nasa.gov/eclipse/solar.html`

Moon

2007 March 3. Total lunar eclipse, visible from eastern parts of North America, South America, Europe, Africa, Asia except far east. Totality starts 22.44 GMT, ends 23.58 GMT.
`http://sunearth.gsfc.nasa.gov/eclipse/lunar.html`

Gemini

Gemini represents the twins of Greek mythology, Castor and Pollux, after whom the constellation's two brightest stars are named. Old star maps show the twins holding hands in the sky. The stars Castor and Pollux mark the heads of the twins. In legend, their mother was Queen Leda of Sparta, but each twin had a different father. Castor was the son of Leda's husband, King Tyndareus, while Pollux was the son of the god Zeus, who visited Leda in the form of a swan. Castor and Pollux were members of the crew of Argonauts who sailed in search of the Golden Fleece. After their death, Zeus set them in the sky. Seafarers regarded the heavenly twins as patron saints, and called upon them for protection in times of danger. The electrical phenomenon known as St Elmo's fire, seen among the masts and rigging of ships, was thought to be the guiding spirit of the twins.

Castor and **Pollux** are contrasting stars. Although they appear in the same region of sky and are of similar brightness, they are not related. **Pollux** is the closer to us, at a distance of 34 light years, while Castor lies 52 light years away. Pollux, also known as Beta (β) Geminorum, is an orange giant of magnitude 1.2, slightly the brighter of the pair.

Castor, Alpha (α) Geminorum, is actually an astounding family of six stars, all linked by gravity. To the naked eye it appears as a single star of magnitude 1.6. But small telescopes, with high magnification, divide Castor into a dazzling pair of blue-white stars of magnitudes 1.9 and 3.0. These orbit each other every 450 years or so. Castor was, in fact, the first pair of stars recognized to be in mutual orbit, a discovery announced by William Herschel in 1803. Also present, but less easy to see in small telescopes, is a 9th-magnitude red dwarf companion some distance from the brighter pair of stars.

All three stars are themselves close doubles, making Castor a stellar sextuplet. Each pair is so close together that the stars cannot be seen separately through even the largest telescopes. Astronomers discovered that the stars were double by analysing their light through a spectroscope, when the presence of light from two stars was recognized. Hence the stars are known as *spectroscopic binaries*.

Castor A, the brightest of the components, consists of two stars bigger and brighter than the Sun orbiting each other every nine days. Castor B, the second-brightest component, consists of another two stars larger than the Sun, orbiting each other every three days. The third component, Castor C, consists of two red dwarfs, smaller and fainter than the Sun, which orbit each other in less than a day. They also eclipse each other, causing the brightness of Castor C to vary from magnitude 9.2 to 9.6. All the

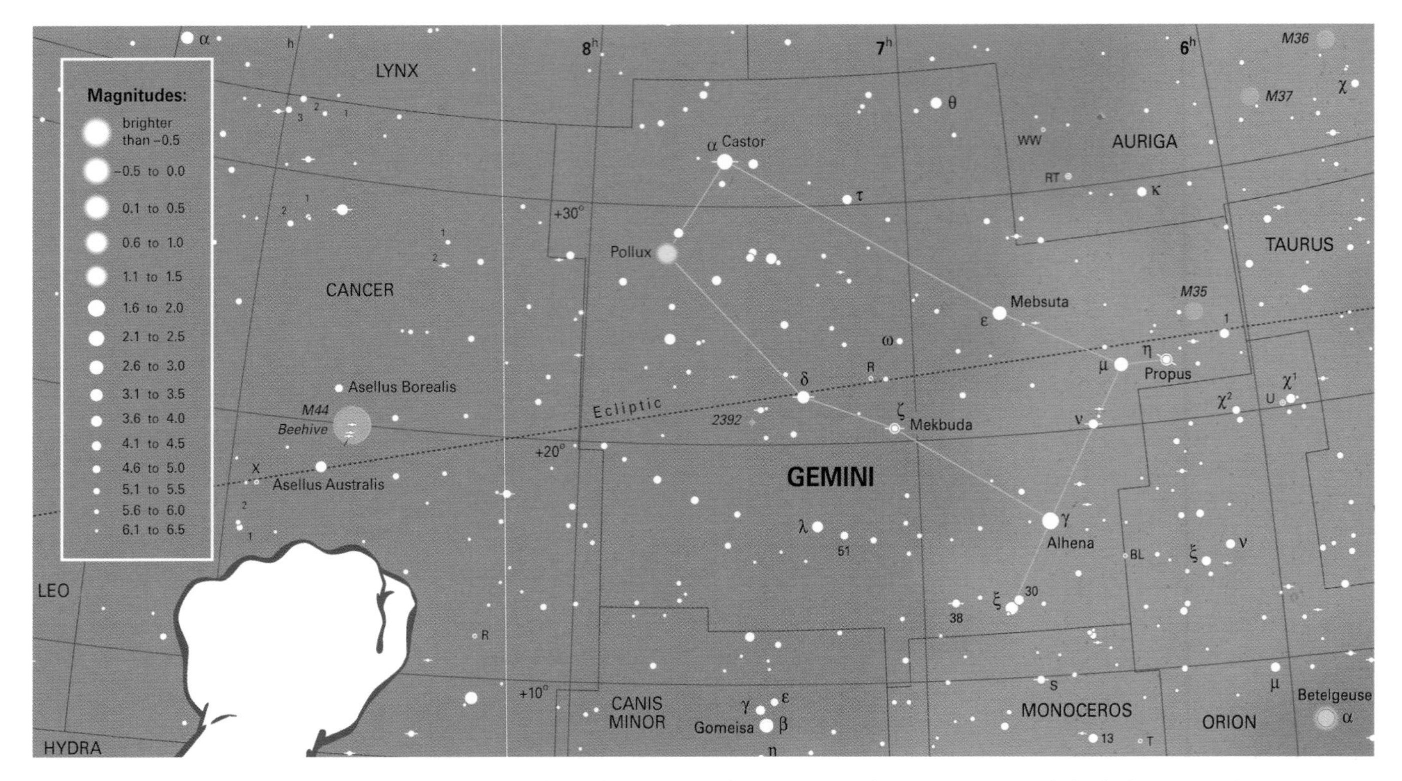

members of the Castor family were born together from the same cloud of gas, and have remained linked by gravity ever since. Try to imagine the view from a planet around one of those stars as you gaze at Castor through a telescope.

Gemini is the home of a large and impressive star cluster, **M35**, near the feet of the twins, close to the border with Taurus. M35 contains over 100 member stars scattered across an area as large as the full Moon. M35 is a fine sight in binoculars, appearing as a mottled patch enveloped in mist. Small telescopes clearly show individual stars, of 8th magnitude and fainter. Unlike many clusters, M35 is not condensed at the centre. Rather, users of telescopes will note that its stars are arranged in disconnected chains, somewhat reminiscent of the lights on a Christmas tree. M35 lies 2800 light years away.

More challenging is the planetary nebula **NGC 2392**. Small telescopes show it as a fuzzy, bluish disk like an 8th-magnitude star out of focus, of similar size to the globe of the planet Saturn. Here, we are seeing a shell of gas thrown off by a dying star, which appears of 10th magnitude at the centre of the nebula. Through large telescopes, NGC 2392 looks like a face surrounded by a fringe; this has given rise to its two popular names, the **Eskimo Nebula** and the **Clown Face Nebula**. Although you will not see such detail through a small telescope, NGC 2392 is one of the easier planetary nebulae to spot, and is well worth searching for.

The Beehive cluster

Between Gemini and Leo lies the constellation of Cancer, the Crab. It is faint, with no star brighter than magnitude 3.5, but at its centre lies a classic object for binoculars, the star cluster **M44**. This cluster was known in ancient times as **Praesepe**, the **Manger**, but it is now popularly termed the **Beehive**. The cluster is flanked by two stars, known as the **Asses**, visualized as feeding at the Manger. To the north is Gamma (γ) Cancri, magnitude 4.7, called Asellus Borealis, the Northern Ass, while Delta (δ) Cancri, magnitude 3.9, is the Southern Ass, Asellus Australis. M44 is visible to the naked eye on clear nights as a misty patch, and was familiar to the ancient Greeks, who kept an eye on it for the purpose of weather forecasting. If the Manger was invisible on an otherwise clear night, this was said to be the sign of a forthcoming storm.

M44 covers 1½° of sky, three times the apparent size of the full Moon – too large to fit into the field of view of a normal telescope but ideal for binoculars. The Beehive cluster is larger even than the Pleiades in Taurus, although its stars are not so bright. The brightest member of the Beehive is Epsilon (ε) Cancri, a white star of magnitude 6.3. In all, M44 contains about 20 stars brighter than 8th magnitude, and several dozen more that are brighter than 10th magnitude. Binoculars show the cluster as a breathtaking sight that resembles a swarm of bees around a hive. This stellar beehive is about 580 light years away.

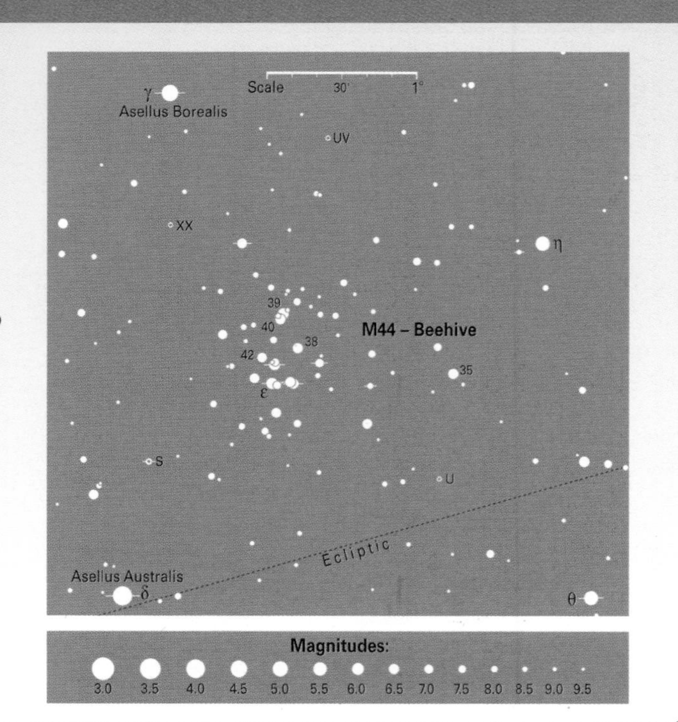

Leo

One of the few constellations that genuinely resembles its name is Leo, the Lion. Legend identifies it with the lion slain by Hercules as one of his twelve labours. The Lion's head is represented by a sickle-shape of six stars, like a back-to-front question mark. The body is outlined by four stars, the tail being marked by Beta (β) Leonis, a white star of magnitude 2.1 whose name, Denebola, comes from the Arabic meaning 'lion's tail'.

At the bottom of the sickle is the brightest star in Leo, Alpha (α) Leonis, better known as **Regulus**, meaning 'little king', appropriate for the king of beasts. Regulus appears in our skies as magnitude 1.4. It is a blue-white star 77 light years away, with an estimated diameter three times that of the Sun, and it gives out over 100 times as much light as the Sun. Small telescopes or even binoculars show that Regulus has a wide companion star of 8th magnitude. Although 700,000 million km (430,000 million miles) from Regulus, more than 100 times the distance of Pluto from the Sun, this companion is genuinely related, and moves through space with Regulus.

The real showpiece of the constellation is the second-brightest star in the sickle, **Gamma** (γ) **Leonis**, also known as Algieba. To the naked eye Gamma Leonis shines at magnitude 2.0. Binoculars show a 5th-magnitude star nearby, 40 Leonis, which is not related but which lies by chance in the same line of sight. Turn a small telescope on to Gamma Leonis and switch to an eyepiece that magnifies about 100 times. It splits into a pair of golden yellow stars, one of the most magnificent doubles in the sky. Both components of Gamma Leonis are orange giants, magnitudes 2.4 and 3.5. They orbit their common centre of mass every 600 years or so. Gamma Leonis is 125 light years away.

Leo contains a number of galaxies, although none are prominent in small instruments. Between Theta (θ) and Iota (ι) Leonis lie the brightest of the Leo galaxies, **M65** and **M66**, both of 9th magnitude. They are spiral galaxies, but M65 is tilted sideways to us and so appears elliptical. In small instruments, M65 and M66 appear as misty patches of light.

If you succeed in finding these, try for the more difficult **M95** and **M96** under the body of Leo. M95 is a barred spiral of 10th magnitude, whereas M96 is an ordinary spiral of 9th magnitude. The structure of each galaxy will not be noticeable in small instruments, for only the brightest central region of each galaxy is visible. All four galaxies lie about 20 million light years away.

As with all such faint, hazy objects their visibility depends critically on viewing conditions. Under poor skies they will be invisible in even moderate-sized telescopes, but under the best conditions they may be glimpsed in binoculars.

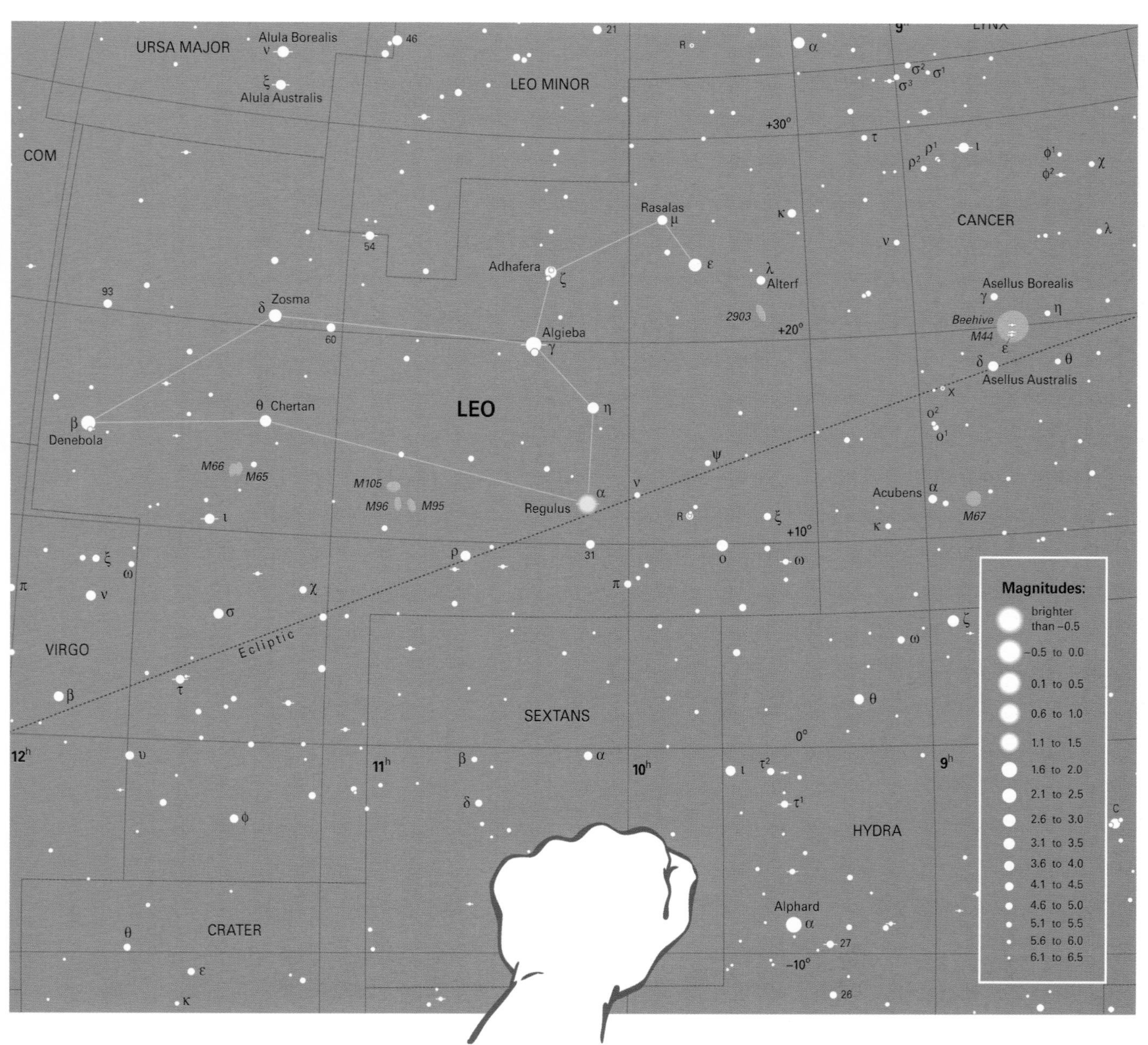

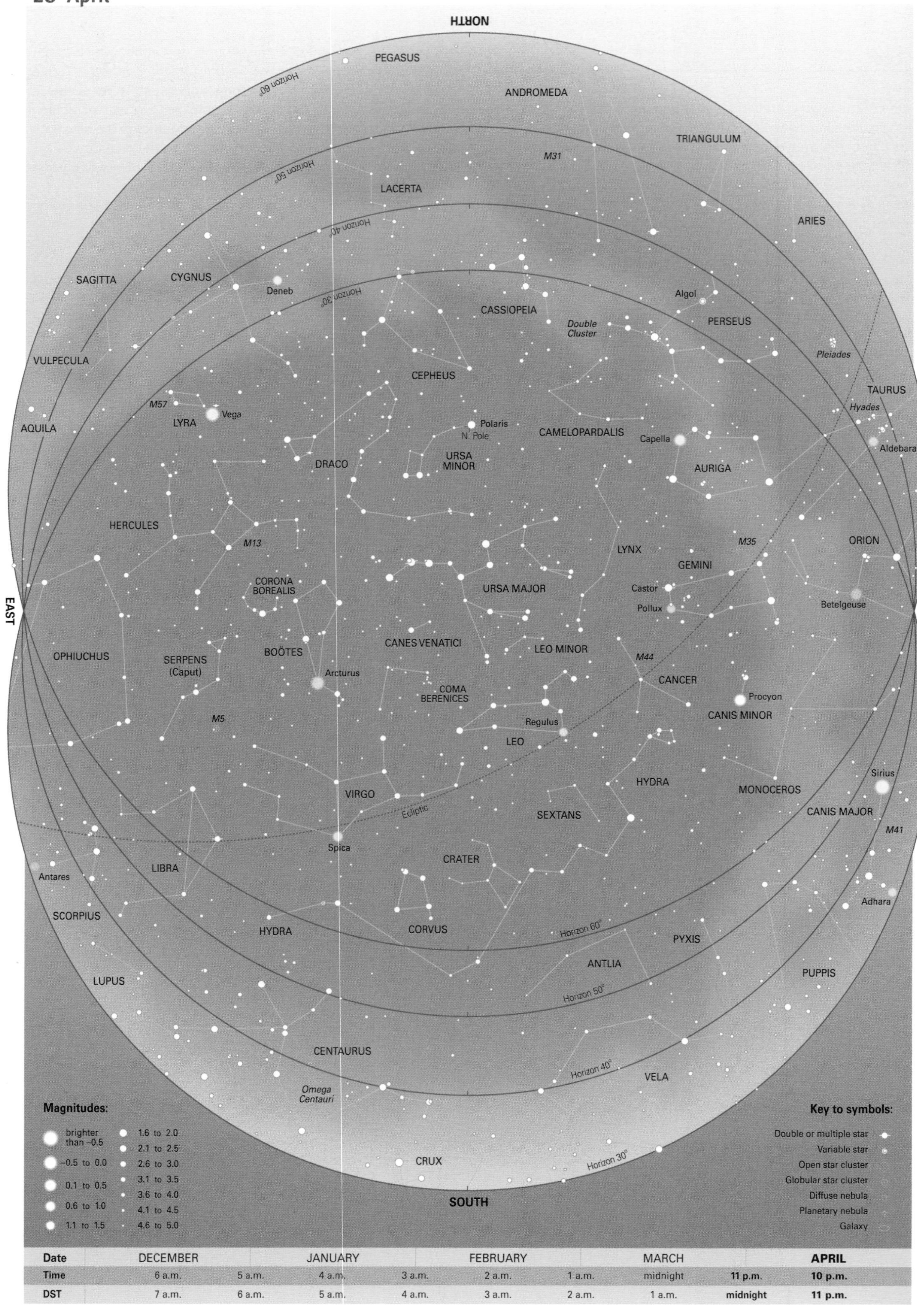

Date	DECEMBER		JANUARY		FEBRUARY		MARCH		**APRIL**
Time	6 a.m.	5 a.m.	4 a.m.	3 a.m.	2 a.m.	1 a.m.	midnight	**11 p.m.**	**10 p.m.**
DST	7 a.m.	6 a.m.	5 a.m.	4 a.m.	3 a.m.	2 a.m.	1 a.m.	**midnight**	**11 p.m.**

April

Key Stars

The stars of the Winter Triangle are becoming swallowed in the evening twilight as the days lengthen. The Plough, or Big Dipper, is best placed for observation this month, being overhead at 10 p.m. Follow the curving handle of the Plough or Dipper to find brilliant Arcturus, and continue the curve to Spica in the south-east. Regulus remains high in the south-west this month while Vega, one of the stars of the Summer Triangle, makes its appearance in the north-east.

The planets this month

Venus

2007 Prominent in the western sky after sunset at mag. –4.1.
2008 Too close to the Sun for observation throughout the month.
2009 Emerges into the eastern dawn sky at month's end.
2010 Low in the western evening twilight at mag. –3.9. Fainter Mercury is below right during the first week or so.
2011 Low in the south-east dawn sky at mag. –3.9, becoming lost in twilight as seen from more northerly latitudes.

Mars

2007 Low in the eastern dawn sky, moving through Aquarius at mag. 1.0.
2008 A first-magnitude evening object in Gemini.
2009 Very low in the eastern twilight before sunrise at mag. 1.2. Ends the month below much brighter Venus.
2010 An evening object in Cancer, passing just north of the Beehive cluster in mid-month at mag. +0.4.
2011 Too close to the Sun to be seen.

Jupiter

2007 Stationary in southern Ophiuchus at mag. –2.4.
2008 Prominent in the morning sky in Sagittarius at mag. –2.2.
2009 In the south-east dawn sky at mag. –2.2.
2010 Emerges from dawn twilight low in the east in the second half of the month, as seen from more southerly latitudes.
2011 At conjunction (behind the Sun) on April 6 and too close to the Sun for observation throughout the month.

Saturn

2007 Stationary on the border between Leo and Cancer at mag. 0.3.
2008 Visible for most of the night in Leo, to the left of Regulus, at mag. 0.5.
2009 Visible for most of the night under the body of Leo at mag. 0.7.
2010 Visible for most of the night between Virgo and Leo at mag. 0.7.
2011 Visible all night in Virgo at mag 0.4. At opposition (due south at midnight) on April 4, 1289 million km (801 million miles) from Earth.

April meteors

One of the year's lesser meteor showers, the **Lyrids**, makes its appearance this month. At best about a dozen meteors are visible each hour, radiating from a point on the border between Lyra and Hercules, near the bright star Vega. Although not numerous, the Lyrids are impressive. A typical Lyrid is bright, fast and leaves a luminous train. Maximum occurs on April 21 or 22 – the exact date varies from year to year – with activity declining rapidly a day or so either side of maximum. The Lyrids are one of the oldest known meteor showers. Recorded sightings go back more than 2000 years, when the shower was much more abundant. Their parent body is Comet Thatcher, which has the longest orbital period of any comet associated with a meteor shower, 415 years.

Ursa Major

The two bears, Ursa Major and Ursa Minor, stand high in the sky on spring evenings. Ursa Major's seven brightest stars comprise one of the most easily recognized patterns in the sky, popularly called the **Plough** or **Big Dipper**. Myths and legends concerning this star group stretch back to the earliest recorded times. We know the constellation as the Great Bear, although it is a strange-looking bear, with a long tail. In Greek mythology, Ursa Major represents the nymph Callisto, who was seduced by Zeus and was subsequently set among the stars in the form of a bear. One version of the tale identifies the Little Bear as her son, Arcas. Another story from ancient Greece says that they are both she-bears, representing the nymphs Adrasteia and Io who raised Zeus when he was an infant.

Ursa Major is the third-largest constellation in the sky. The seven stars that make up the familiar saucepan shape of the **Plough** or **Dipper** are only part of the complete constellation, forming the rump and tail of the bear. All the stars of the

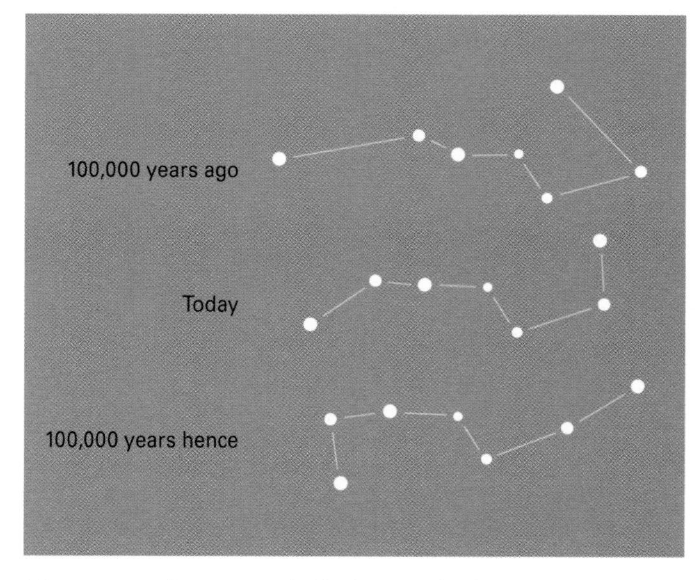

All stars are on the move, gradually changing the shapes of constellations. Here, the stars of the Plough are shown as they appeared 100,000 years ago, as they are today, and as they will appear 100,000 years hence.

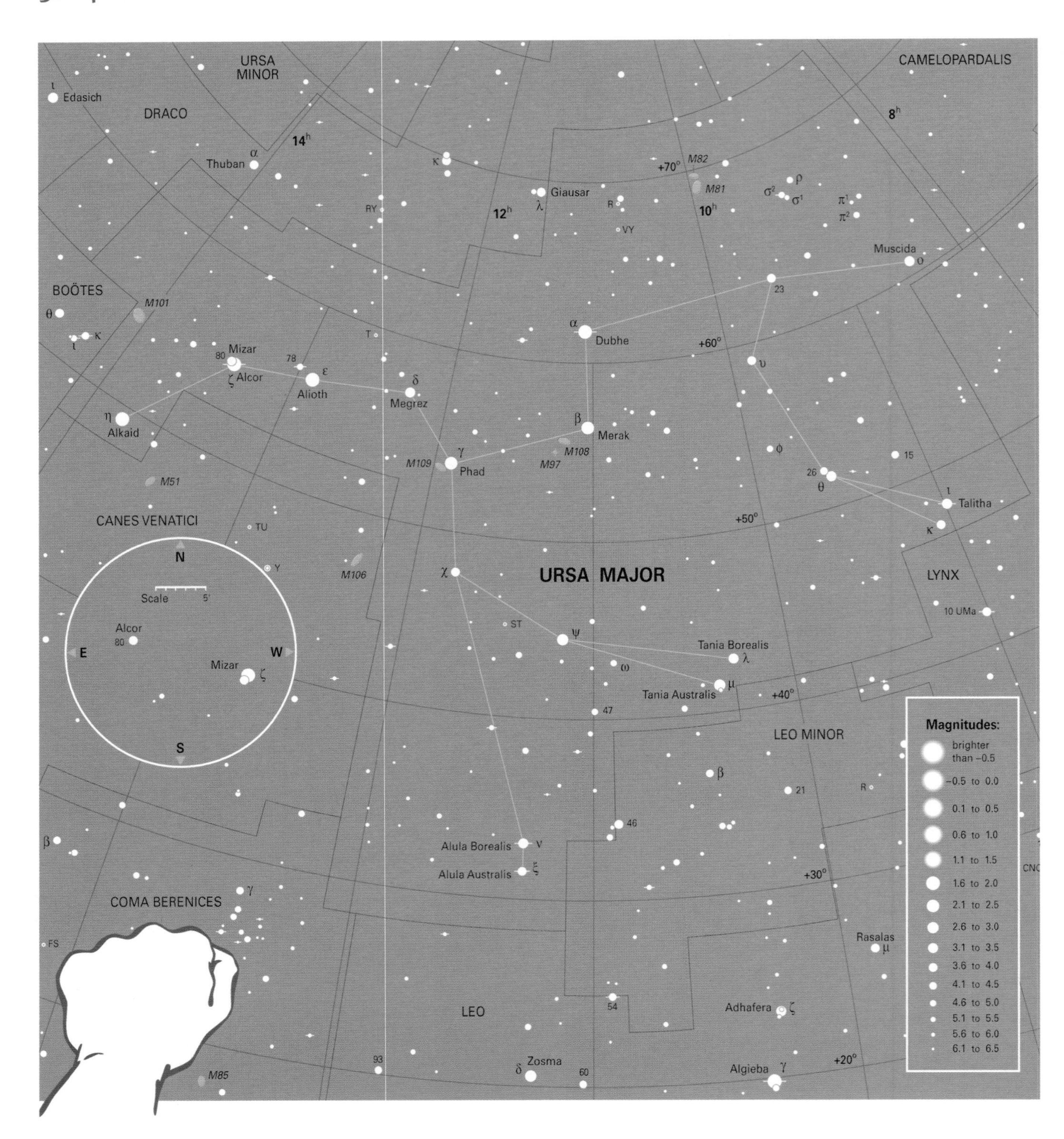

saucepan are of 2nd magnitude except Delta (δ) Ursae Majoris, which is 3rd magnitude.

Unlike most constellations, in which the brightest stars have no connection with each other, five of the seven stars in the saucepan are related, moving at the same speed and in the same direction through space. These five stars were evidently born together, and now form a very scattered cluster. The two non-members of the cluster are Alkaid (also known as Benetnasch), at the end of the handle, and Dubhe, in the bowl. Over long periods of time, the motions of Alkaid and Dubhe relative to the other five stars considerably alter the familiar saucepan shape (*see diagram on previous page*).

The most celebrated star of Ursa Major is situated at the bend of the saucepan's handle, second from the end. It is **Zeta** (ζ) **Ursae Majoris**, popularly known as **Mizar**, magnitude 2.2. Keen eyesight shows a 4th-magnitude star, **Alcor**, nearby. Mizar and Alcor are sometimes termed the horse and rider. They both lie about 80 light years away, but are probably too far apart to form a genuine binary. However, they are moving through space together, along with the other members of the Ursa Major cluster.

If you look at Mizar through a small telescope you will see that it also has a much closer companion of 4th magnitude that is definitely associated with it. Precise observations over more than two centuries have confirmed that this star is very slowly moving around Mizar on an orbit that takes about 10,000 years to complete.

This close companion to Mizar was discovered in 1650 by the Italian astronomer Giovanni Riccioli. It made Mizar the first double star to be discovered with a telescope, and it was one of

The Whirlpool galaxy

Just over the border from Ursa Major, in the constellation of Canes Venatici, the Hunting Dogs, lies **M51**, a celebrated spiral galaxy popularly known as the **Whirlpool**. It was the first galaxy in which spiral structure was noted, by Lord Rosse in 1845. A remarkable feature of the Whirlpool is the satellite galaxy at the end of one of its spiral arms, as pictured in long-exposure photographs. M51, of 8th magnitude, can be glimpsed under good conditions in binoculars as a misty patch. Small telescopes should show the starlike central nuclei of the Whirlpool and its satellite, but large telescopes are needed to trace its spiral arms. The Whirlpool Galaxy lies about 26 million light years from us.

several 'firsts' for Mizar. Mizar was also the first double star to be photographed, by George P. Bond at Harvard in 1867; it was the first star discovered to be a spectroscopic binary, by Edward C. Pickering at Harvard in 1889; and it is often the first double star that amateur astronomers turn to with their telescopes. A spectroscopic binary is a pair of stars that are too close together to be seen separately in a telescope, and whose double nature can be inferred only from its spectrum. Actually, Mizar's 4th-magnitude companion is also a spectroscopic binary, making Mizar a four-star family. In addition, Alcor turns out to be another spectroscopic binary, completing a remarkable stellar grouping.

Another famous double, although far more challenging, is **Xi** (ξ) **Ursae Majoris**. It was the first double star to have its orbit calculated, in 1828, a demonstration that the same laws of gravity that govern the orbits of the planets around the Sun also apply to the stars. The two components, of magnitudes 4.3 and 4.8, orbit each other every 60 years, and the movement is so rapid that it can be followed from year to year in amateur telescopes. When the stars were closest together, around 1993, a telescope of 150 mm (6 inches) aperture was needed to separate them, but since 1998 they have been divisible with apertures of only 75 mm (3 inches). By 2017 they will be within range of 60-mm (2.4-inch) telescopes. The stars will reach their widest separation in about the year 2035.

Ursa Major is the home of several galaxies that lie far outside our Milky Way. In the northern part of the constellation lie M81 and M82, a prominent duo. **M81** is one of the most beautiful spiral galaxies and, at 7th magnitude, one of the easiest to see in small instruments. It appears as a milky-white blur almost half the apparent diameter of the full Moon, much brighter towards its centre and somewhat elliptical in shape because it is tilted at an angle to us. Under clear, dark skies you can pick it out in binoculars. M81 is similar to the great spiral galaxy in Andromeda, M31, although not as large.

Half a degree to the north of M81, and hence visible in the same low-power field of view, is an altogether more peculiar galaxy, **M82**, which appears cigar-shaped. M82 is smaller than M81 and is only one-quarter as bright. For many years M82 was thought to be an exploding galaxy, but more recent research has concluded that it is actually a spiral galaxy that was disturbed after brushing past M81 about 600 million years ago. It appears elongated because we see it edge-on. M82 and M81 both lie about 12 million light years from us.

Another major galaxy in Ursa Major is **M101**, a face-on spiral around 22 million light years away. It lies north of the end of the Big Dipper's handle, forming a triangle with Alcor and Alkaid. Through binoculars and small telescopes M101 appears as a large, pale disk. Clear, dark skies are needed to see it.

Ursa Minor

The Little Bear, Ursa Minor, looks indeed like a smaller but fainter version of Ursa Major. Its stars form a saucepan shape reminiscent of the Big Dipper, but with the curve of its handle reversed. Unlike Ursa Major, the stars of the Little Dipper are not related. Two stars in the Little Dipper's bowl, **Beta** (β) **Ursae Minoris, Kochab**, of 2nd magnitude, and **Gamma** (γ) **Ursae Minoris, Pherkad**, of 3rd magnitude, are known as the **Guardians of the Pole**. Keen eyesight, or binoculars, shows that Pherkad has a wide 5th-magnitude companion, which is actually an unrelated foreground star. Note also that the faintest star in the bowl, **Eta** (η) **Ursae Minoris**, consists of two widely spaced 5th-magnitude stars, also unrelated.

The most famous star in Ursa Minor is Alpha (α) Ursae Minoris, better known as **Polaris**, the Pole Star. It lies less than a degree (two Moon diameters) from the celestial north pole, so it is directly overhead to an observer at the Earth's north pole. Although it shines modestly at magnitude 2.0 as seen from Earth, were we to travel the 430 light years to its vicinity we would find that it is in reality a most imposing star – a yellow supergiant perhaps 100 times the Sun's diameter and giving out as much light as 2500 Suns. Through a telescope, Polaris appears like a pearl strung on a necklace of fainter surrounding stars.

Telescopes also reveal that Polaris is a double star with a 9th-magnitude companion, although the companion is difficult to detect in the smallest apertures because it is so much fainter than Polaris. The companion slowly orbits Polaris over many thousands of years. Polaris is the remarkable case of a Cepheid variable that has almost stopped pulsating. Its pulsations progressively decreased during the twentieth century, stabilizing in the 1990s at a few hundredths of a magnitude – an example of a star evolving as astronomers watched. It is not known whether Polaris' pulsations will one day start to build up again.

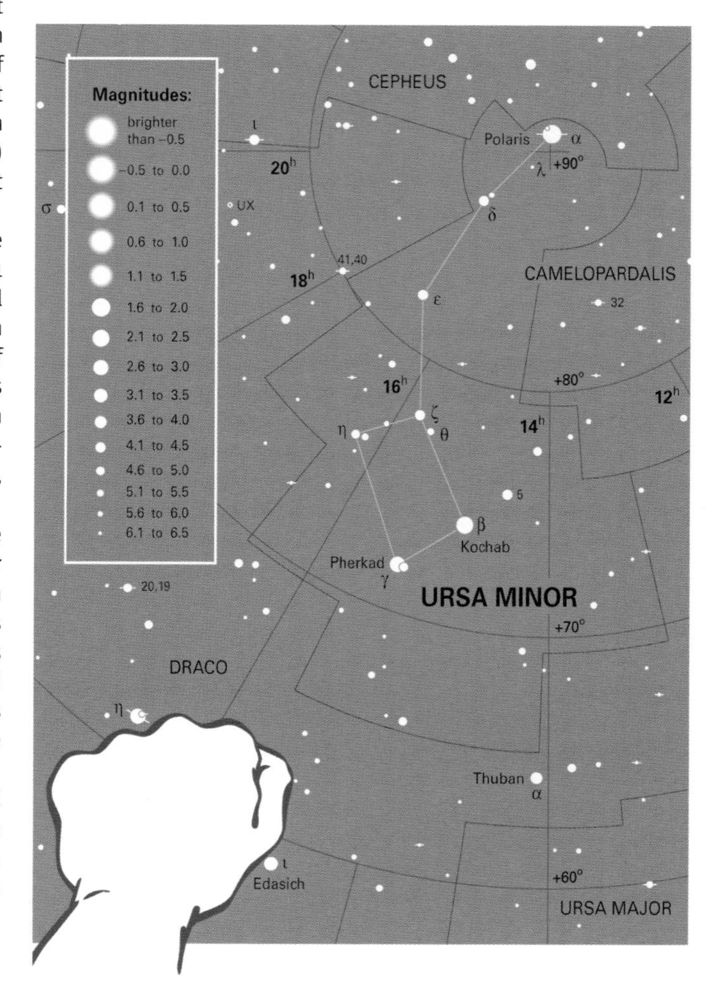

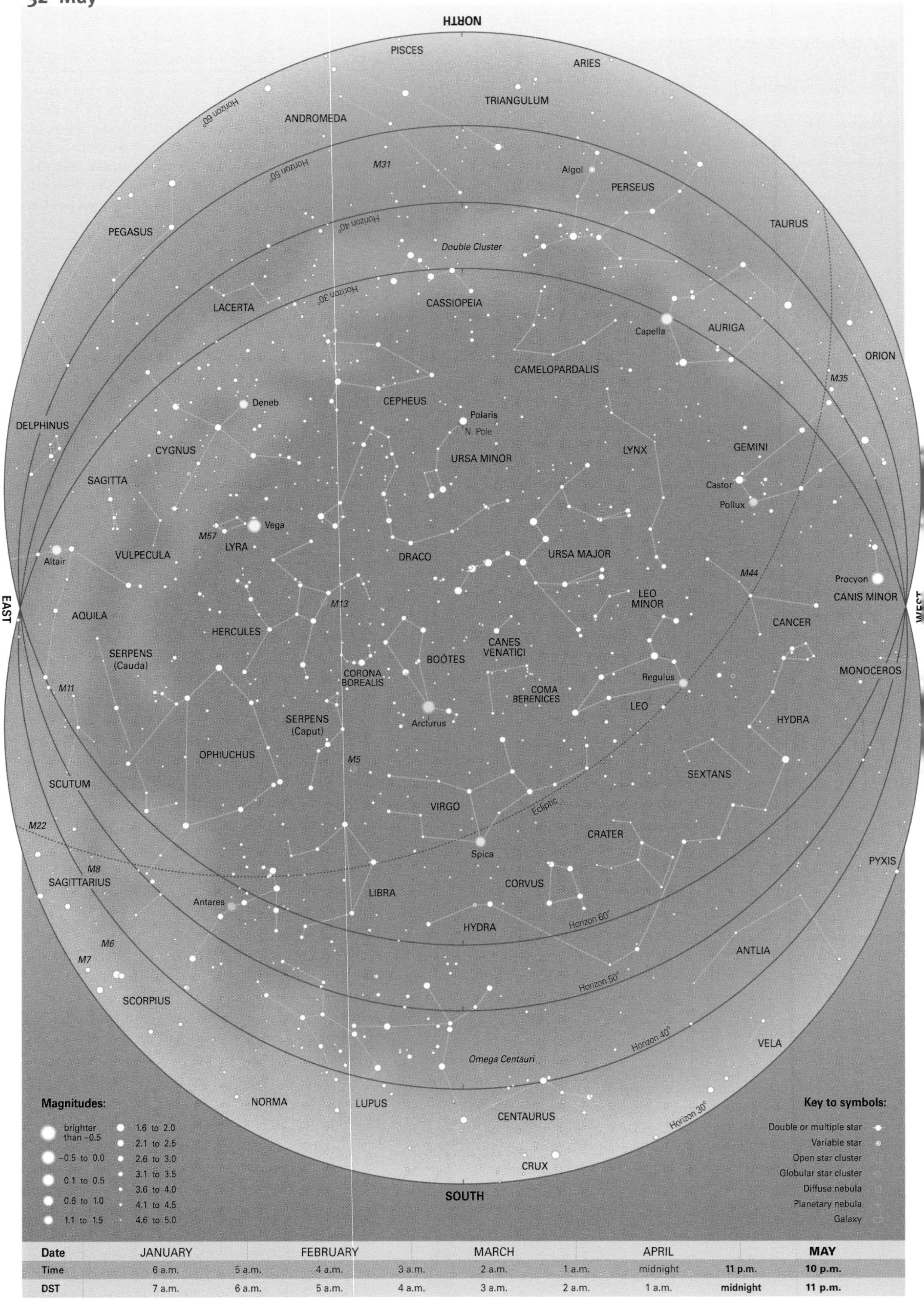

NORTH

PISCES
ARIES
TRIANGULUM
ANDROMEDA
Horizon 60°
Horizon 50°
Horizon 40°
Horizon 30°
M31
Algol
PERSEUS
PEGASUS
TAURUS
Double Cluster
LACERTA
CASSIOPEIA
Capella
AURIGA
CAMELOPARDALIS
ORION
M35
DELPHINUS
Deneb
CEPHEUS
Polaris
N. Pole
URSA MINOR
LYNX
GEMINI
CYGNUS
SAGITTA
Castor
Pollux
Vega
M57
LYRA
Altair
VULPECULA
DRACO
URSA MAJOR
M44
Procyon
CANIS MINOR
EAST
WEST
M13
LEO MINOR
AQUILA
HERCULES
CANCER
SERPENS (Cauda)
CANES VENATICI
BOÖTES
CORONA BOREALIS
Regulus
MONOCEROS
M11
COMA BERENICES
LEO
SERPENS (Caput)
Arcturus
HYDRA
OPHIUCHUS
M5
SEXTANS
SCUTUM
VIRGO
Ecliptic
M22
CRATER
Spica
PYXIS
M8
SAGITTARIUS
LIBRA
CORVUS
Antares
HYDRA
Horizon 60°
M6
ANTLIA
M7
Horizon 50°
SCORPIUS
VELA
Horizon 40°
Omega Centauri
NORMA
LUPUS
CENTAURUS
Horizon 30°
CRUX
SOUTH

Magnitudes:

- brighter than −0.5
- −0.5 to 0.0
- 0.1 to 0.5
- 0.6 to 1.0
- 1.1 to 1.5
- 1.6 to 2.0
- 2.1 to 2.5
- 2.6 to 3.0
- 3.1 to 3.5
- 3.6 to 4.0
- 4.1 to 4.5
- 4.6 to 5.0

Key to symbols:

- Double or multiple star
- Variable star
- Open star cluster
- Globular star cluster
- Diffuse nebula
- Planetary nebula
- Galaxy

Date	JANUARY		FEBRUARY		MARCH		APRIL		MAY
Time	6 a.m.	5 a.m.	4 a.m.	3 a.m.	2 a.m.	1 a.m.	midnight	11 p.m.	10 p.m.
DST	7 a.m.	6 a.m.	5 a.m.	4 a.m.	3 a.m.	2 a.m.	1 a.m.	midnight	11 p.m.

May

Key Stars

Orange Arcturus blazes high overhead, while Spica glints in the south. Castor and Pollux are sinking towards the western horizon, followed by Regulus and the stars of Leo. Ursa Major still stands high in the north. In the east, the stars of the Summer Triangle, led by blue-white Vega, are rising. Ruddy Antares is climbing into view in the south-east, although it is still low on the horizon for observers above 50° north. For more southerly observers, this is the best month to sweep the rich starfields of Centaurus and Lupus, now well above the southern horizon.

The planets this month

Venus

2007 Prominent in the western evening sky among the stars of Gemini at mag. −4.2.
2008 Too close to the Sun to be seen.
2009 Low in the eastern dawn sky throughout the month at mag. −4.4.
2010 In the western evening sky at mag. −3.9.
2011 Swamped by twilight in the eastern dawn sky throughout the month.

Mars

2007 A first-magnitude object low in the eastern dawn sky in Pisces.
2008 An evening object in Cancer at mag. 1.3, passing in front of the Beehive Cluster around May 22.
2009 Low in the eastern dawn twilight at mag. 1.2, below left of brilliant Venus.
2010 A first-magnitude object in the western evening sky, moving from Cancer into Leo.
2011 Emerges into the eastern dawn sky by the end of the month, as seen from more southerly latitudes.

Jupiter

2007 Visible for most of the night in southern Ophiuchus at a prominent mag. −2.6.
2008 Visible prominently for much of the night at mag. −2.5, stationary in Sagittarius.
2009 A prominent morning object in Capricornus at mag. −2.4.
2010 Low in the south-east dawn sky at mag. −2.2.
2011 Emerges into the eastern dawn sky in the second half of the month for more southerly observers.

Saturn

2007 On the border between Leo and Cancer at mag. 0.5.
2008 Stationary In Leo, to the left of Regulus, at mag. 0.6.
2009 An evening object, stationary under the body of Leo at mag. 0.9.
2010 An evening object between Virgo and Leo at mag. 0.9.
2011 Visible for most of the night in Virgo at mag. 0.7.

May meteors

Dust from Halley's Comet can be seen burning up in the atmosphere this month, producing the **Eta** (η) **Aquarid** meteor shower. At the shower's best, around May 5–6 each year, about 35 swift-moving meteors per hour radiate from near the star Eta (η) Aquarii, in the northern part of Aquarius. The shower's peak is broad, and activity remains high for several days either side of maximum. But the shower is difficult to see from the northern hemisphere, for two reasons. Since the radiant lies virtually on the celestial equator, observers need to be south of latitude 40° north to have much hope of seeing any Eta Aquarids at all. Additionally, the radiant does not rise until nearly 2 a.m., and from northern latitudes does not get very high in the sky before dawn, which considerably restricts observations. In all, this is a difficult shower to observe.

Boötes

High in the spring sky stands the large constellation of Boötes, the Herdsman. It is easily found, lying next to the tail of Ursa Major. In legend, Boötes and Ursa Major are inextricably linked. In some tales Boötes is visualized as a man ploughing or driving a wagon (Ursa Major was often visualized as a wagon as well as a plough), while in other legends he is a herdsman or hunter chasing the Great Bear around the pole.

The constellation's brightest star is **Arcturus**, a name which is Greek for 'bear keeper'. It was an important star to the ancient Greeks, who used its rising and setting as a guide to the changing seasons. Nowadays, the reappearance of Arcturus in the early evening is a welcome sign of approaching spring.

At magnitude −0.05, Arcturus is the fourth-brightest star in the entire sky, marginally brighter than Vega in Lyra (page 41). Look carefully and you will see that it has a noticeably orange tint, which is more prominent when viewed through binoculars. Arcturus is the type of star known as a red giant. It is red (or more strictly orange) because it has a relatively cool surface temperature, two-thirds that of the Sun. And it is indeed a giant, with a diameter 27 times that of the Sun. Arcturus gives out over 100 times as much light as the Sun, and it lies 37 light years away.

Although Arcturus is so much bigger and brighter than the Sun, the two stars are actually quite similar in mass. So why do they appear so different? The reason is that Arcturus is nearing the end of its life while the Sun is still in middle age. As stars age, they swell up in size to become red giants. One day our Sun will turn into a red giant, although fortunately not for several thousand million years yet. So when you look at Arcturus, reflect that you are seeing a preview of the Sun as it will be in 5000 million years' time. When the Sun becomes a red giant, the Earth and all life on it will be roasted to a cinder. Perhaps Arcturus once had a life-bearing planet.

Boötes is well stocked with attractive double stars, but the most celebrated of them is very difficult to divide in small amateur telescopes because the components are so close together. The star concerned is **Epsilon** (ε) **Boötis**, sometimes known as Izar, 'the girdle', or Pulcherrima, meaning 'most beautiful' because of the contrasting colours of its stars. To the naked eye it appears of magnitude 2.4, but actually it consists of an orange giant of magnitude 2.5 and a greenish-blue companion of magnitude 4.7. They are 2.9 seconds of arc apart, which would be within the power of a 50-mm (2-inch) refractor to split if they were of similar brightness. However, in practice the brightness difference means that the fainter star is difficult to separate from the primary's glare with even a 75-mm (3-inch) telescope. High magnification and very steady air are needed to divide the two, but the pair provide a stunning sight to those who succeed.

As an easier task, turn to **Kappa** (κ) **Boötis** in the north of the constellation, consisting of 5th- and 7th-magnitude white stars easily divisible in small telescopes. A similar but closer pair is **Pi** (π) **Boötis**, in the south of the constellation. For a warmer-toned pair, find **Xi** (ξ) **Boötis**, a 5th-magnitude golden-yellow star with a 7th-magnitude orange companion that orbits it every 150 years. Their distinct hues make them a showpiece duo.

More complex is the magnitude 4.3 white star **Mu** (μ) **Boötis**, whose name Alkalurops comes from the Arabic referring to the Herdsman's crook or staff. Binoculars show that it has a 7th-magnitude companion. But this companion is itself a close double whose components orbit each other every 250 years. These components can be split in telescopes of 75 mm (3 inches) aperture with high magnification.

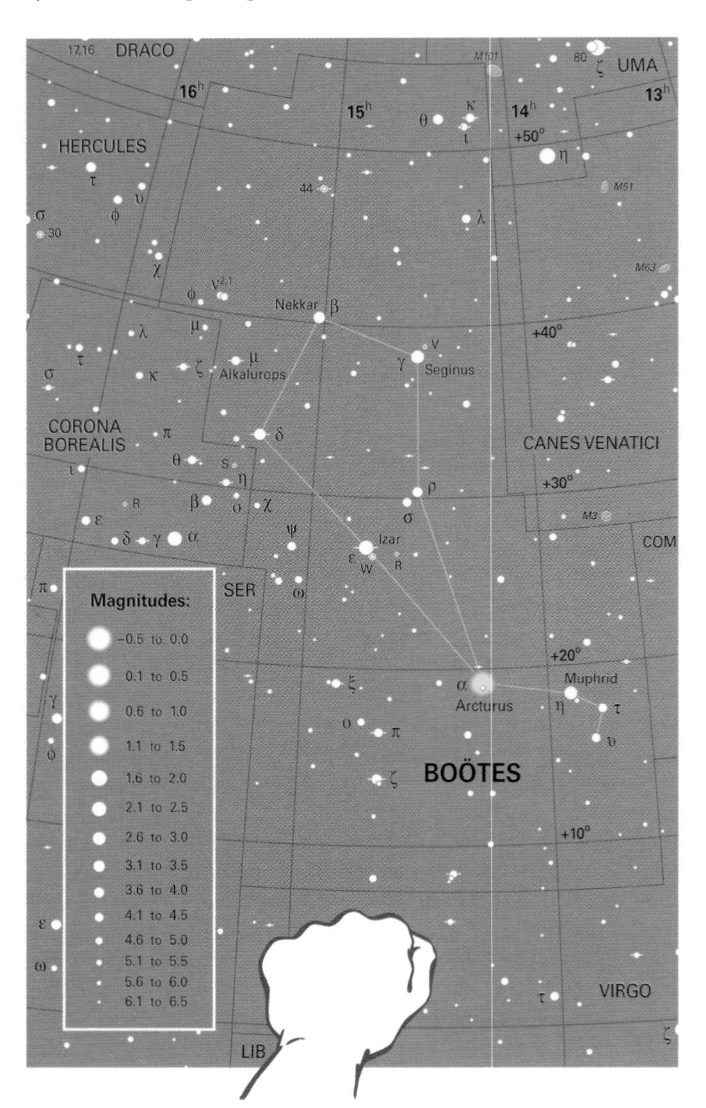

Virgo

In the south around 10 p.m. this month lies the second-largest of all the constellations: Virgo, the Virgin. She is usually depicted as a maiden holding an ear of wheat, represented by the bright star Spica. As such she is identified with Demeter, the harvest goddess. But Virgo has a second identity, that of the goddess of justice. In this guise she is depicted holding the scales of justice, represented by the neighbouring constellation of Libra.

Despite its large size Virgo is not a particularly prominent constellation, with the notable exception of blue-white **Spica**, magnitude 1.0. Spica is actually over twice as hot as Sirius and 100 times more luminous. But in our skies Spica appears fainter than Sirius because of its considerably greater distance, 260 light years.

Virgo contains one of the most celebrated double stars, **Gamma** (γ) **Virginis**, also called Porrima after a Roman goddess of prophecy. To the naked eye Gamma Virginis appears as a star of magnitude 2.7, but it actually consists of a pair of identical white stars, each of magnitude 3.5, which orbit each other every 169 years. Their orbital motion is noticeable in small telescopes. The two stars were closest together in the year 2005, when they were indivisible in apertures below about 250 mm (10 inches), but are now moving apart again and will become an easy object for small telescopes after 2012, remaining so for the rest of the century.

Virgo's secret treasure is a cluster of distant galaxies, which spills across its northern border into Coma Berenices. The **Virgo Cluster** contains about 3000 galaxies, centred on the giant elliptical galaxy **M87**, which is visible as a 9th-magnitude smudge in small telescopes or even binoculars. M87 is known to radio astronomers as the powerful radio source **Virgo A**. Long-exposure photographs show a jet of luminous gas being shot out of M87, as though the galaxy has suffered some violent event. Astronomers think that the activity in M87 is due to a massive black hole at its centre which ejects gas along its axis of rotation, accounting for the visible jet.

The Virgo Cluster is of particular interest to astronomers because it is the nearest large cluster of galaxies to us. Most galaxies, perhaps all, are grouped together in clusters of various sizes. Our own Galaxy is a member of a cluster of three dozen members called the Local Group, and according to some astronomers the Local Group is itself a member of a so-called supercluster centred on the Virgo Cluster. The distance of the Virgo Cluster has been measured by the Hubble Space Telescope as 50 million light years. At this distance none of the galaxies appears particularly large or bright, but a number of them are within range of small telescopes. In addition to M87, mentioned above, look for **M49**, an 8th-magnitude elliptical galaxy; **M60**, a 9th-magnitude elliptical galaxy; **M84** and **M86**, a pair of 9th-magnitude elliptical galaxies.

Another galaxy of note in Virgo is found on the constellation's southern border with Corvus. This is **M104**, an 8th-magnitude spiral galaxy that is not a member of the Virgo Cluster but somewhat closer to us, 30 million light years away. We see M104 edge-on, so although it is actually a spiral it appears elliptical through small telescopes. It is popularly known as the **Sombrero Galaxy**, because it resembles a wide-brimmed hat on long-exposure photographs, but do not expect to see it as anything more than an elongated smudge through a small telescope.

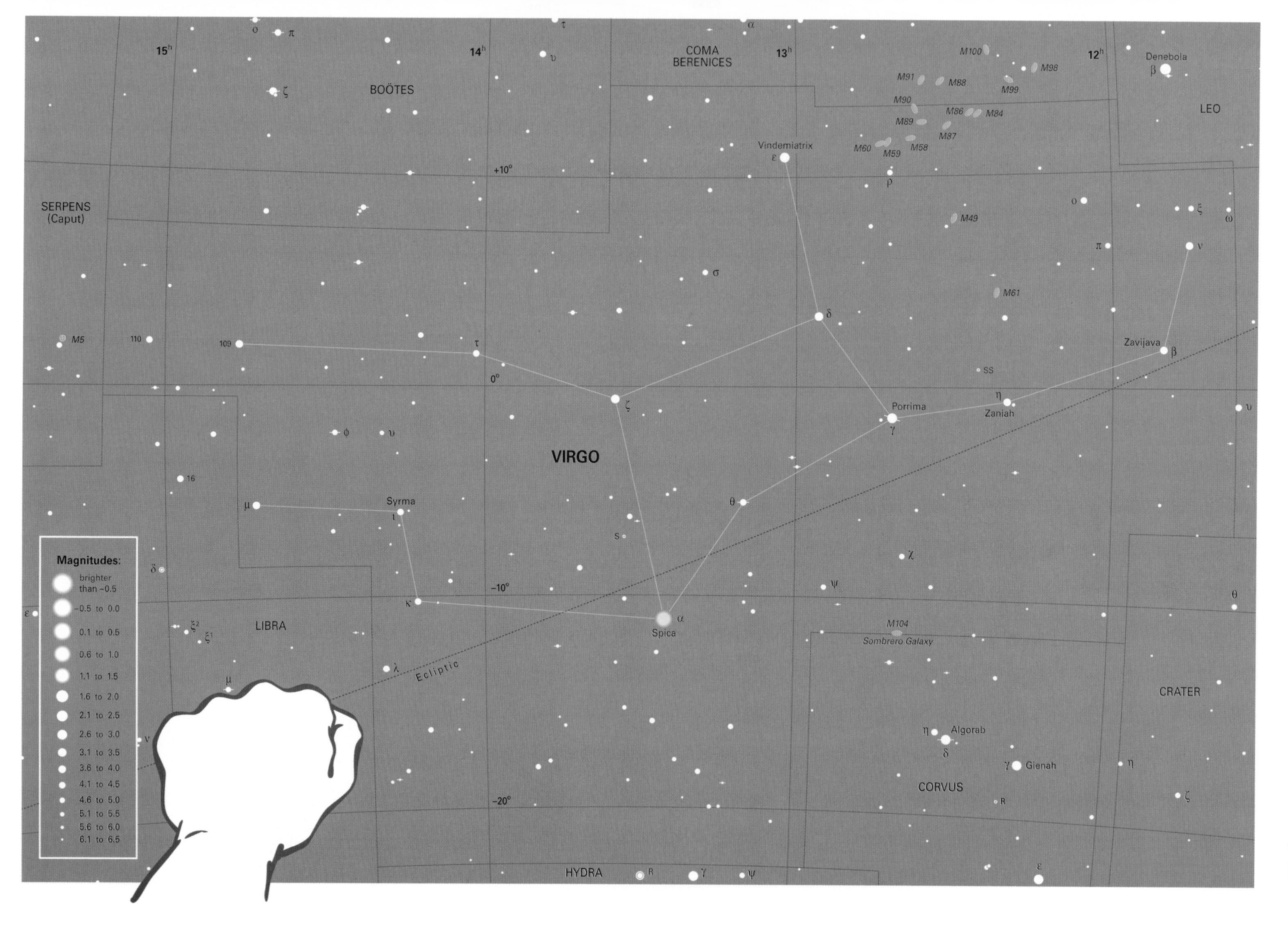
BOÖTES
COMA BERENICES
LEO
Denebola
SERPENS (Caput)
Vindemiatrix
VIRGO
Porrima
Zaniah
Zavijava
Syrma
Spica
LIBRA
Ecliptic
M104
Sombrero Galaxy
CRATER
Algorab
Gienah
CORVUS
HYDRA
M5
M49
M61
M100
M98
M91
M88
M99
M90
M86
M84
M89
M87
M60
M59
M58
Magnitudes:
brighter than –0.5
–0.5 to 0.0
0.1 to 0.5
0.6 to 1.0
1.1 to 1.5
1.6 to 2.0
2.1 to 2.5
2.6 to 3.0
3.1 to 3.5
3.6 to 4.0
4.1 to 4.5
4.6 to 5.0
5.1 to 5.5
5.6 to 6.0
6.1 to 6.5

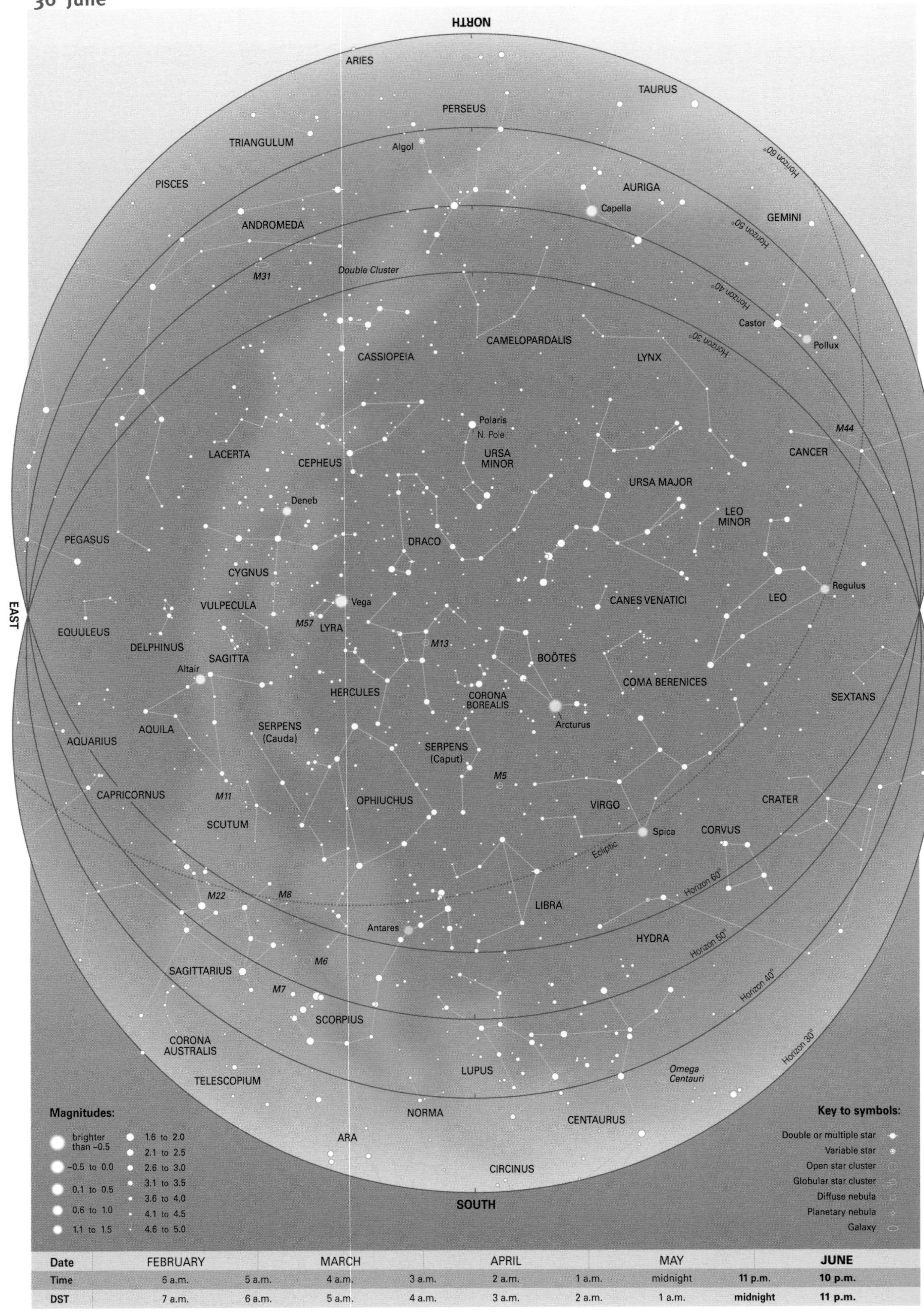

Date	FEBRUARY		MARCH		APRIL		MAY		JUNE
Time	6 a.m.	5 a.m.	4 a.m.	3 a.m.	2 a.m.	1 a.m.	midnight	**11 p.m.**	**10 p.m.**
DST	7 a.m.	6 a.m.	5 a.m.	4 a.m.	3 a.m.	2 a.m.	1 a.m.	**midnight**	**11 p.m.**

June

Key Stars

Arcturus sparkles overhead as the sky darkens on June evenings. Spica is still prominent in the south-west, although Regulus and the stars of Leo are closer to the western horizon and vanish before midnight. In the north, the Big Dipper (or Plough) stands on its handle. In the south are Antares and the stars of Scorpius, brushing the horizon for observers in high northern latitudes but well displayed for those farther south. Above Antares is the sparse region occupied by the large constellations Ophiuchus and Serpens. Becoming prominent in the eastern sky are Vega, Deneb and Altair, the stars of the Summer Triangle.

The planets this month

Venus

2007 Prominent in the western evening sky at mag. −4.4. At greatest elongation (maximum separation) from the Sun of 45° on June 9. Ends the month near fainter Saturn.
2008 At superior conjunction (directly behind the Sun) on June 9 and invisible throughout the month.
2009 Low in the eastern dawn sky at mag. −4.2 close to fainter Mars. Greatest elongation (maximum separation) from the Sun of 46° is on June 5.
2010 In the western evening twilight at mag. −4.0.
2011 Too close to the Sun to be seen throughout the month.

Mars

2007 In the eastern dawn sky in Pisces at mag. 0.8.
2008 In the western evening sky, moving from Cancer into Leo at mag. 1.6. Ends the month next to Regulus.
2009 Low in the eastern dawn twilight at mag. 1.1, passing brilliant Venus.
2010 In the western evening sky in Leo at mag. 1.2, passing north of Regulus at the end of the first week.
2011 In the eastern dawn sky at mag. 1.4, but very low as seen from more northerly latitudes.

Jupiter

2007 Visible all night in southern Ophiuchus at mag. −2.6. Opposition (due south at midnight) on June 5, 644 million km (400 million miles) from Earth.
2008 Visible all night in Sagittarius at a prominent mag. −2.6.
2009 Prominent in the morning sky on the boundary between Capricornus and Aquarius at mag. −2.6.
2010 A prominent morning object in Pisces at mag. −2.4.
2011 In the eastern dawn twilight at mag. −2.2.

Saturn

2007 In the western evening sky at mag. 0.6. Ends the month close to brilliant Venus.
2008 An evening object in Leo at mag. 0.8.
2009 In the south-west evening sky under the body of Leo at mag. 1.0.
2010 A first-magnitude object in the western evening sky between Virgo and Leo.
2011 An evening object, stationary in Virgo at mag. 0.8.

Eclipses

Sun

2011 June 1. Partial solar eclipse, maximum 60%, visible from northern Alaska, northern Canada and north-east Asia.
http://sunearth.gsfc.nasa.gov/eclipse/SEcat/SEdecade2011.html

Moon

2011 June 15. Total lunar eclipse, visible from Europe, Africa, Asia, Australasia, eastern South America. Totality starts 19.22 GMT, ends 21.03 GMT.
http://sunearth.gsfc.nasa.gov/eclipse/LEcat/LEdecade2011.html

Hercules

Two giants stand head-to-head in the June sky. Hercules represents the Greek hero who undertook twelve labours on the orders of King Eurystheus of Mycenae. The other figure is Ophiuchus, a man entwined by a serpent. Both constellations are large – Hercules is the fifth-largest in the sky and Ophiuchus the eleventh-largest – but neither of them is particularly prominent. Hercules contains no star brighter than 3rd magnitude and consequently is not easy to find. Look for him between the bright stars Vega and Arcturus.

Old star maps depict Hercules as a man on one knee, brandishing a club. One foot, marked by Iota (ι) Herculis, is resting on the head of Draco, the Dragon. To the ancient Greeks this constellation was simply The Kneeling Man, 'whose name none can tell, nor what he labours at', as the Greek poet Aratos wrote in the third century BC. The identification with Hercules came later. The fact that this constellation was anonymous to the ancient Greeks suggests that they did not originate it, but inherited it from an earlier civilization, probably the Babylonians of the Middle East.

Hercules appears upside-down in the sky. His head lies in the south of the constellation, near the border with Ophiuchus, and is marked by the star **Alpha** (α) **Herculis**, popularly known as **Rasalgethi**, from the Arabic meaning 'kneeler's head'. Alpha Herculis is usually classed as a red giant, but sometimes it is put among the supergiants. In common with most giant stars it changes in size, varying in brightness as it does so. It fluctuates between 3rd and 4th magnitude with no set period.

Alpha Herculis is one of the largest stars known, but as with all such distended stars the exact dimensions remain somewhat uncertain. At its distance of nearly 400 light years it must emit as much light as 1000 Suns and have a diameter a few hundred times that of the Sun, so huge that it could easily engulf the orbit of the planet Mars.

Alpha Herculis is also a glorious but tight double star for small telescopes. Its close companion of 5th magnitude appears greenish by contrast with the burnished copper colour of Alpha Herculis itself. The two stars form a slow-moving binary, orbiting every few thousand years.

Hercules is a good hunting ground for double stars. Start with **Delta** (δ) **Herculis**, an unrelated duo of 3rd and 8th magnitudes. Another unrelated pair is **Kappa** (κ) **Herculis**, a 5th-magnitude yellow giant with a wide 6th-magnitude orange giant companion. **Rho** (ρ) **Herculis** is a neat 5th- and 6th-magnitude pair for small telescopes, both white in colour. Look next at **95 Herculis**, an attractive pair of 5th-magnitude stars appearing silver and gold in colour like Christmas-tree decorations. Finally, do not miss **100 Herculis**, a matched pair of 6th-magnitude white stars, easily divisible by small telescopes.

The most distinctive part of Hercules is the **'keystone'** representing his pelvis, formed by four stars: Zeta (ζ), Eta (η), Epsilon (ε) and Pi (π) Herculis. Between Eta and Zeta Herculis lies one of the showpieces of the heavens, the great globular cluster **M13**, a city of 300,000 stars or more, compressed into a ball 100 light years across.

M13 is one of 150 globular clusters that are scattered in a halo around our Galaxy. It is the brightest globular cluster in northern skies, glimpsed by the naked eye on clear nights as a luminous patch resembling a faint star out of focus. Binoculars show it as a mysterious glowing ball, covering about one-third the diameter of the full Moon and with two faint stars standing like sentinels either side. Note how its brightness increases towards the centre, where the stars are most crowded. The brightest stars in the cluster are giants, emitting the light of a thousand Suns. Telescopes of 100 mm (4 inches) aperture and above show those giants superimposed on the general background glow, giving the cluster an amazing granular appearance at high magnification. Anyone who lived on a planet in M13 would see the sky filled with thousands of dazzling stars, some rivalling the full Moon in brilliance. There would be no true night at all.

Hercules contains a second globular cluster, **M92**, smaller than M13 and about half the brightness but still easily visible through binoculars. Compare the two clusters and note that M92 has a brighter centre than M13, because its stars are more densely packed. This gives it a more stellar appearance than M13 and in fact M92 can at first be mistaken for a star. M92 is nearly 30,000 light years away, somewhat farther than M13.

Ophiuchus

One of the lesser-known constellations is Ophiuchus, which straddles the celestial equator between Hercules and Scorpius. Ophiuchus represents the Greek god of medicine Asclepius (the Roman Aesculapius). His symbol was the snake, and in the sky Ophiuchus is inextricably linked with a serpent, in the form of the constellation Serpens, which is visualized as being wrapped around him. Ophiuchus divides Serpens into two, an arrangement that is unique among the constellations. Serpens Caput, the serpent's head, lies on the right of Ophiuchus as we look at it while Serpens Cauda, the tail, is on the left. Nevertheless, the two halves of Serpens are regarded as one constellation. This celestial triptych of Ophiuchus and the two halves of Serpens can be seen in the south during summer nights.

Ophiuchus contains a number of globular clusters, the brightest being **M10**, **M12** and **M62**. All three, of 7th magnitude, are visible in binoculars, but they pale by comparison with M13 in Hercules. Far more appealing is the open cluster **IC 4665**, a scattering of 20 or so stars of 7th magnitude and fainter, in the same field as the orange giant Beta (β) Ophiuchi. This cluster is ideal for binoculars because of its large apparent size, greater than that of the full Moon.

For an attractive double star, turn a small telescope on **36 Ophiuchi** in the constellation's southern reaches and see a neat pair of matching 5th-magnitude orange dwarf stars. Then look farther north to **70 Ophiuchi**, a beautiful golden-yellow and orange duo, magnitudes 4.2 and 6.0, that orbit each other every 88 years. The stars are currently moving apart as seen from Earth, and are divisible in the smallest telescopes. It should be interesting to watch their separation increase year by year, reaching a maximum around 2025. No observer should miss 70 Ophiuchi, one of the most famous double stars in the sky.

On the constellation's border with Scorpius, just north of Antares, lies an outstanding multiple star for small apertures, **Rho** (ρ) **Ophiuchi**. Low magnification shows a V-shaped grouping consisting of a 5th-magnitude star at the apex, accompanied by two stars of 7th magnitude. High powers show that the star at the triangle's apex is itself double, consisting of a tight pair of 5th and 6th magnitudes. On long-exposure photographs this whole area is seen to be enveloped in a faint haze of nebulosity that extends southwards to Antares.

Barnard's Star

Oddly enough, the most celebrated star in Ophiuchus is not visible to the naked eye. Nor is it easy to find in small telescopes. Yet it is worth hunting for, because it is the second-closest star to the Sun and also the fastest-moving star in our skies.

It is Barnard's Star, a red dwarf of magnitude 9.5, lying 5.9 light years away. This faint star is named after the American astronomer Edward Emerson Barnard who noticed in 1916 that it had changed markedly in position on two photographs taken 22 years apart. All stars are on the move, but most of them move almost imperceptibly. Barnard's Star has the fastest movement across the sky of any star, crossing the apparent diameter of the full Moon in 180 years.

In addition to its movement across the sky, Barnard's Star is also approaching the Sun. It will reach its minimum distance from us in AD 11,800, when it will be 3.8 light years away, closer even than Alpha Centauri is today. Yet it is so feeble that even then it will appear of only magnitude 8.5. Barnard's Star has a mass about 15 per cent that of the Sun, and is 2300 times fainter than the Sun.

Barnard's Star lies to the left of Beta (β) Ophiuchi, near the 5th-magnitude star 66 Ophiuchi, which is slightly variable. The chart shows how rapidly the star's position moves. Observe it over a period of years so that you, too, can detect the motion of this runaway star.

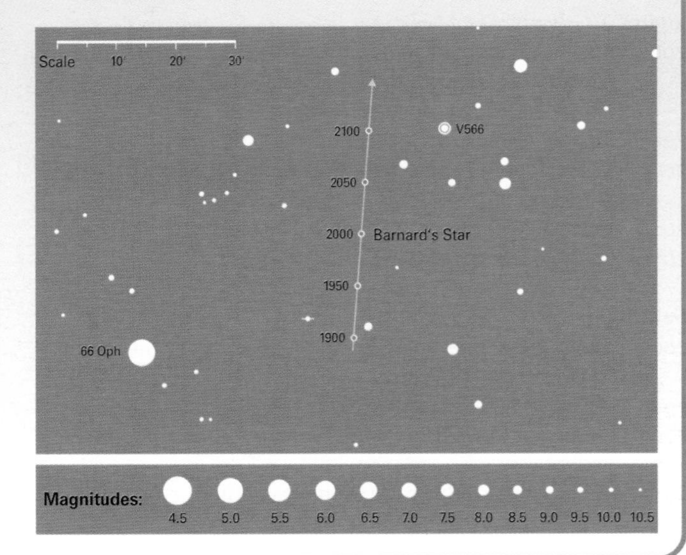

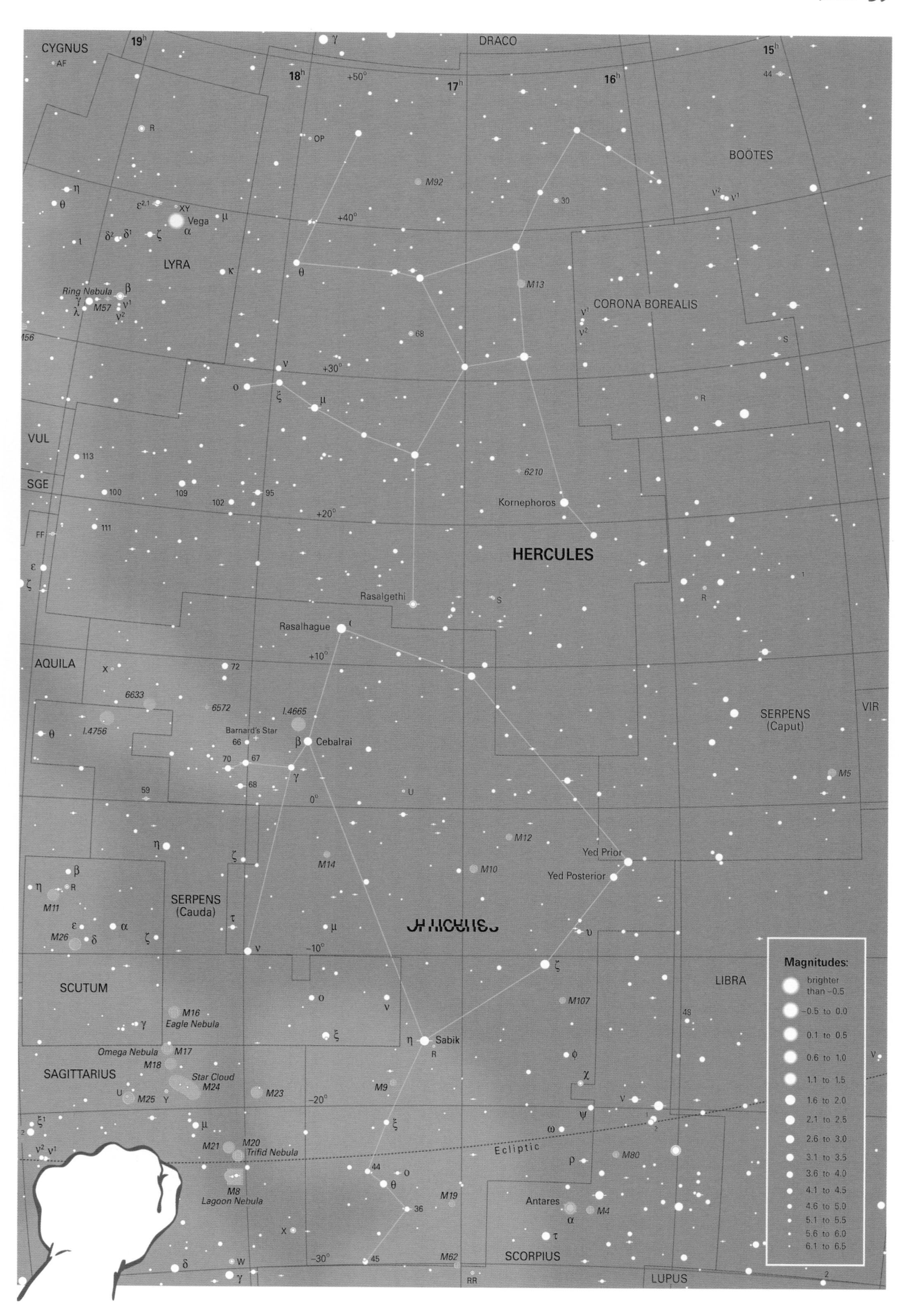

CYGNUS
DRACO
BOÖTES
LYRA
Vega
Ring Nebula
M57
M92
M13
CORONA BOREALIS
HERCULES
Kornephoros
Rasalgethi
Rasalhague
6210
VUL
SGE
AQUILA
6633
6572
I.4665
I.4756
Barnard's Star
Cebalrai
SERPENS
(Caput)
VIR
M5
M12
M14
M10
Yed Prior
Yed Posterior
SERPENS
(Cauda)
M11
M26
SCUTUM
LIBRA
M107
M16
Eagle Nebula
Omega Nebula
M17
M18
SAGITTARIUS
Star Cloud
M24
M25
M23
M9
Sabik
Ecliptic
M80
M21
M20
Trifid Nebula
M8
Lagoon Nebula
M19
Antares
M4
M62
SCORPIUS
LUPUS
Magnitudes:
brighter than −0.5
−0.5 to 0.0
0.1 to 0.5
0.6 to 1.0
1.1 to 1.5
1.6 to 2.0
2.1 to 2.5
2.6 to 3.0
3.1 to 3.5
3.6 to 4.0
4.1 to 4.5
4.6 to 5.0
5.1 to 5.5
5.6 to 6.0
6.1 to 6.5

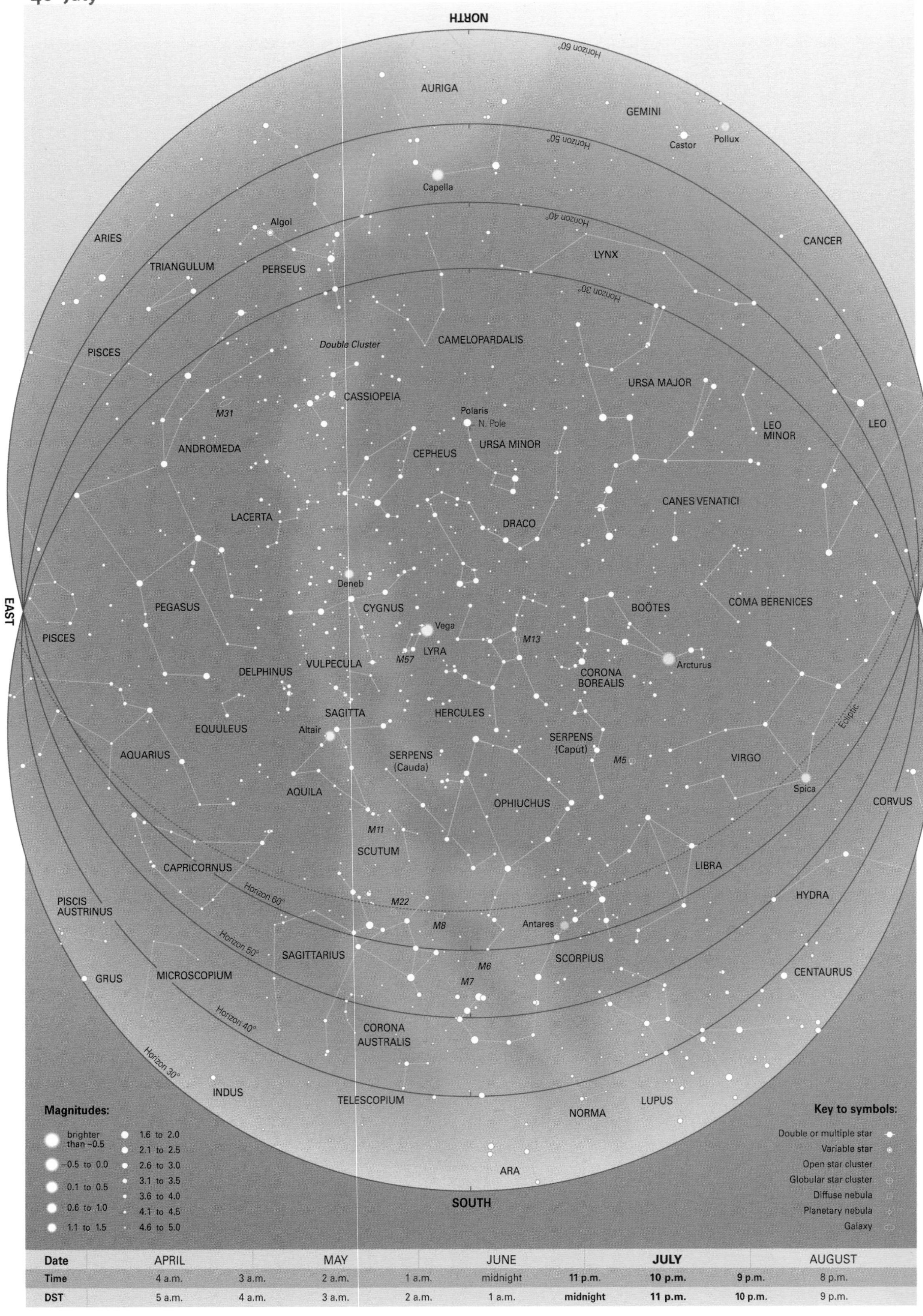

Date	APRIL		MAY		JUNE		**JULY**		AUGUST
Time	4 a.m.	3 a.m.	2 a.m.	1 a.m.	midnight	**11 p.m.**	**10 p.m.**	**9 p.m.**	8 p.m.
DST	5 a.m.	4 a.m.	3 a.m.	2 a.m.	1 a.m.	**midnight**	**11 p.m.**	**10 p.m.**	9 p.m.

July

Key Stars

Brilliant blue-white Vega sparkles like a sapphire overhead, leading the Summer Triangle of stars across the July sky. In the south-west, Arcturus is sinking and Spica is about to vanish below the horizon. Antares can still be seen low in the south-west. The southern starfields of Scorpius and Sagittarius are displayed at their best in the evening sky this month. Capella twinkles low on the northern horizon, as seen from high northerly latitudes. On the eastern horizon the Square of Pegasus heralds the arrival of the stars of autumn.

The planets this month

Venus

2007 Starts the month in the western evening sky near fainter Saturn but becomes swamped by evening twilight after mid-July.
2008 Too close to the Sun to be seen.
2009 In the eastern dawn sky all month at mag. −4.1.
2010 Low in the western evening twilight throughout the month at mag. −4.1.
2011 Too close to the Sun to be seen throughout the month.

Mars

2007 A morning object, moving from Aries into Taurus at mag. 0.5.
2008 Low in the western evening sky, passing brighter Saturn in the second week of July.
2009 A first-magnitude object in the eastern dawn sky, passing between the Pleiades and Hyades clusters in Taurus.
2010 Low in the western evening sky at mag. 1.4, becoming lost in twilight by month's end from more northerly latitudes. Passes below slightly brighter Saturn in the last week of July.
2011 In the eastern dawn sky at mag. 1.4.

Jupiter

2007 A prominent evening object in southern Ophiuchus at mag. −2.5.
2008 Visible all night in Sagittarius at mag. −2.7. At opposition (due south at midnight) on July 9, 622 million km (387 million miles) from Earth.
2009 Visible for most of the night in Capricornus at a prominent mag. −2.8.
2010 A prominent morning object, stationary in Pisces at mag. −2.6.
2011 In the eastern morning sky in Aries at mag. −2.3.

Saturn

2007 Starts the month close to brilliant Venus, low in the western sky after sunset at mag. 0.6, but becomes lost in evening twilight by month's end.
2008 Low in the western evening sky at mag. 0.8, passing fainter Mars in the first half of the month and then becoming lost in twilight to more northerly observers.
2009 A first-magnitude object in the western evening twilight.
2010 Low in the western evening twilight, above fainter Mars.
2011 In the south-west evening sky in Virgo at mag. 0.9.

Eclipses

Sun

2009 July 21-22. Total solar eclipse, visible from China and western Pacific Ocean; maximum duration of totality 6m 39s. A partial eclipse is visible from east and south-east Asia and Hawaii.
http://sunearth.gsfc.nasa.gov/eclipse/solar.html

2010 July 11. Total solar eclipse, visible from south-western South America and southern Pacific Ocean; maximum duration of totality 5m 20s.
http://sunearth.gsfc.nasa.gov/eclipse/solar.html

Lyra

Overhead this month lies the compact and distinctive constellation of Lyra, the Lyre, which represents the harp of the Greek bard Orpheus. In legend, Hermes made the harp from a tortoise shell and gave it to his half-brother Apollo. In turn, Apollo passed the harp to Orpheus who, it was said, could charm wild beasts and even trees with his musical skill. This ability was particularly valuable when Orpheus entered the fearful Underworld to plead for the return of his dead wife, Eurydice. Hades, god of the Underworld, was so moved by the music of Orpheus that he agreed to release Eurydice to the land of the living, on condition that Orpheus did not look back at her as he led her to the upper world. But, just before they emerged into daylight, Orpheus glanced round to be sure she was following, and lost her for ever. Thereafter he roamed the world in sorrow, playing his harp plaintively. After Orpheus died, his harp was placed in the heavens in commemoration of his musical gifts.

The constellation has also been identified as an eagle or vulture, often depicted on early star maps carrying the harp. Whatever your interpretation of Lyra, it has a rich stock of celestial attractions to charm astronomers.

Lyra is easily located by its brightest star Alpha (α) Lyrae, better known as **Vega**, the fifth-brightest star in the entire sky. Vega's name comes from the Arabic meaning 'swooping eagle', for the Arabs were among those who visualized the constellation as an eagle. Vega is a blue-white star of magnitude 0.03, lying 25 light years away. It has a mass of about three Suns and gives out 50 times as much light as the Sun. In 1983 the Infra-Red Astronomical Satellite, IRAS, made the astounding discovery that Vega is surrounded by a disk of cold, dark dust that is forming into a system of planets – another solar system. Vega is a much younger star than the Sun, with an age of only about 350 million years. Perhaps, in billions of years, life will evolve on one of those planets that are now forming around Vega.

Near to Vega, binoculars or even sharp eyesight shows a wide 5th-magnitude double star, **Epsilon** (ε) **Lyrae**, which has a delightful surprise awaiting the users of telescopes. Both stars are themselves double, forming a spectacular stellar quadruplet that is popularly known as the **Double Double**.

Epsilon-1, the slightly wider pair, consists of stars of magnitudes 5.0 and 6.1 orbiting every 1750 years or so. Currently they are closing together and will continue to do so through the twenty-first century, although only slowly. The current separation is 2.4 arc seconds. Stars with such separations should in theory be divisible by an aperture of 50 mm (2 inches), but the brightness difference means that in practice at least 60 mm (2.4 inches) will be required.

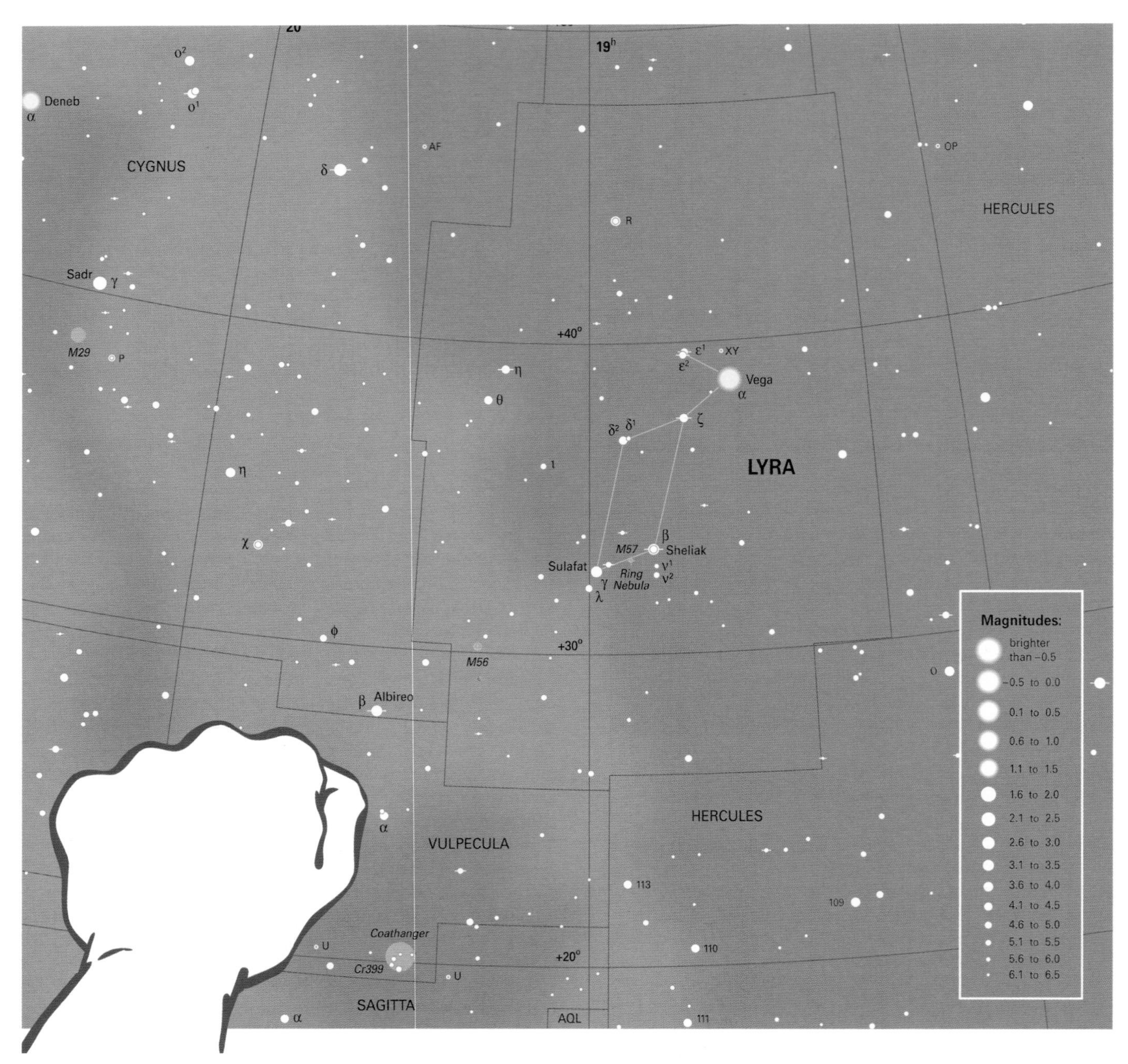

Epsilon-2 Lyrae, the closer pair, consists of stars of magnitudes 5.2 and 5.5 whose orbit lasts over 700 years. Their separation is 2.4 arc seconds, and will also be difficult to split in 50-mm (2-inch) apertures. But whatever your telescope, do not miss this showpiece object, the finest quadruple star in the sky.

Epsilon Lyrae is not the only double star in this constellation with a surprise up its sleeve. Look at **Beta** (β) **Lyrae**, an attractive coupling of a cream primary with an 8th-magnitude blue star, easily separated by small telescopes. But the primary star is itself a complex eclipsing binary that varies between magnitudes 3.3 and 4.4 every 13 days. Compare its brightness with that of Gamma (γ) Lyrae, constant at magnitude 3.2. Astronomers have struggled for decades to understand the curious primary of Beta Lyrae, whose period of variation is gradually lengthening. The answer has come from carefully studying the star's light through spectroscopes. Evidently the two eclipsing stars are so close together (closer than Mercury is to the Sun) that each is distorted into an egg shape by the other's gravity. Gas flows between the stars, some of it spiralling away into space. When you look at Beta Lyrae, try to imagine the turbulent activity taking place there, unseen by the human eye.

Another double-variable is **Delta** (δ) **Lyrae**, an easy binocular duo consisting of a 4th-magnitude red giant star that is slightly variable and a blue-white star of magnitude 5.6. The two lie at different distances from us and so are unrelated. Look also at **Zeta** (ζ) **Lyrae**, a binocular pair of magnitudes 4.3 and 5.7.

Between Beta and Gamma Lyrae lies another of the treasures of Lyra, the famous **Ring Nebula, M57**, a celestial smoke ring thrown off by a dying star about 2000 light years away. M57 is a type of object known as a planetary nebula, not because it has anything to do with planets but because it shows a rounded disk like a planet when viewed through a telescope. Such an object is believed to be formed when a star like the Sun reaches the end of its life. After swelling up into a red giant, the star sloughs off its distended outer layers to form a planetary nebula, leaving the core as a faint white dwarf. Observations by the Hubble Space Telescope have revealed that the Ring Nebula is not really a ring at all but a cylinder of gas which we happen to view end-on.

Small telescopes show M57 as a faint hazy spot, larger than the disk of Jupiter. Apertures of at least 100 mm (4 inches) are needed to discern its ring shape and its elliptical outline. Its central star, of 15th magnitude, is too faint for amateur telescopes. Unfortunately, M57 is not as impressive visually as it is on long-exposure photographs.

Vulpecula

Adjoining Lyra is Vulpecula, the Fox, a little-known constellation well worth an introduction. Like its southerly neighbour Sagitta, Vulpecula straddles a rich area of the Milky Way and is a good hunting ground for novae. The constellation was placed in the sky in 1690 by the Polish astronomer Johannes Hevelius under the title of Vulpecula cum Anser, the Fox and Goose, which has since been shortened. Hevelius said that he placed a fox here to be close to two other predatory animals, the vulture (Lyra) and the eagle (Aquila). Its brightest star, **Alpha** (α) **Vulpeculae**, magnitude 4.4, has an unrelated companion of 6th magnitude easily visible in binoculars.

Vulpecula achieved unexpected fame in 1967 when radio astronomers at Cambridge, UK, discovered the first of the flashing radio stars called pulsars south of Alpha Vulpeculae. Pulsars are thought to be the highly condensed cores of stars that have exploded as supernovae. Like most pulsars, this one is so faint that it is invisible optically.

Close to the border with Sagitta lies one of the oddest-looking star groupings in the sky, sometimes known either as **Brocchi's Cluster** or **Collinder 399** but more commonly termed the **Coathanger** because of its distinctive shape. Its main feature is an almost perfectly straight line of six stars, from the centre of which protrudes a curving hook of four more stars, completing the coathanger shape. The ten stars of the Coathanger are from 5th to 7th magnitude and the whole object is 1½° long, making it an ideal binocular sight.

However, the Coathanger is not a true cluster, because its stars lie at a wide range of distances – from just over 200 light years to more than 1000 light years – and are all moving through space in different directions. Hence the stars cannot be related, and lie in the same line of sight only by chance. Despite this, the Coathanger remains a delightful surprise.

From the Coathanger, turn your binoculars to the best-known object in Vulpecula, **M27**, popularly known as the **Dumbbell Nebula**. M27 is a planetary nebula, a shell of gas thrown off by a dying star, like the Ring Nebula in Lyra. But the Ring Nebula and the Dumbbell have noticeably different appearances.

Whereas many planetary nebulae are difficult to spot because they are so small, the Dumbbell can be missed because it is unexpectedly large. At 8th magnitude the Dumbbell is bright enough to show up in binoculars as a smudgy patch even under indifferent sky conditions. It is elongated, with a maximum diameter about a quarter that of the full Moon, eight times larger than the Ring Nebula. Having a favourable combination of size and brightness, M27 is the most conspicuous of all planetary nebulae, at least in small instruments.

Seen through a telescope, the Dumbbell is greenish in colour. It gets its name from its double-lobed shape in telescopes, reminiscent of an hour-glass, or perhaps a figure eight, in distinct contrast to the rounded shape of the Ring Nebula. Its central star, the white dwarf that has shed its outer layers to form the nebula, is of 14th magnitude, too faint for small telescopes. The Dumbbell lies about 1000 light years away, among the closest planetary nebulae to us.

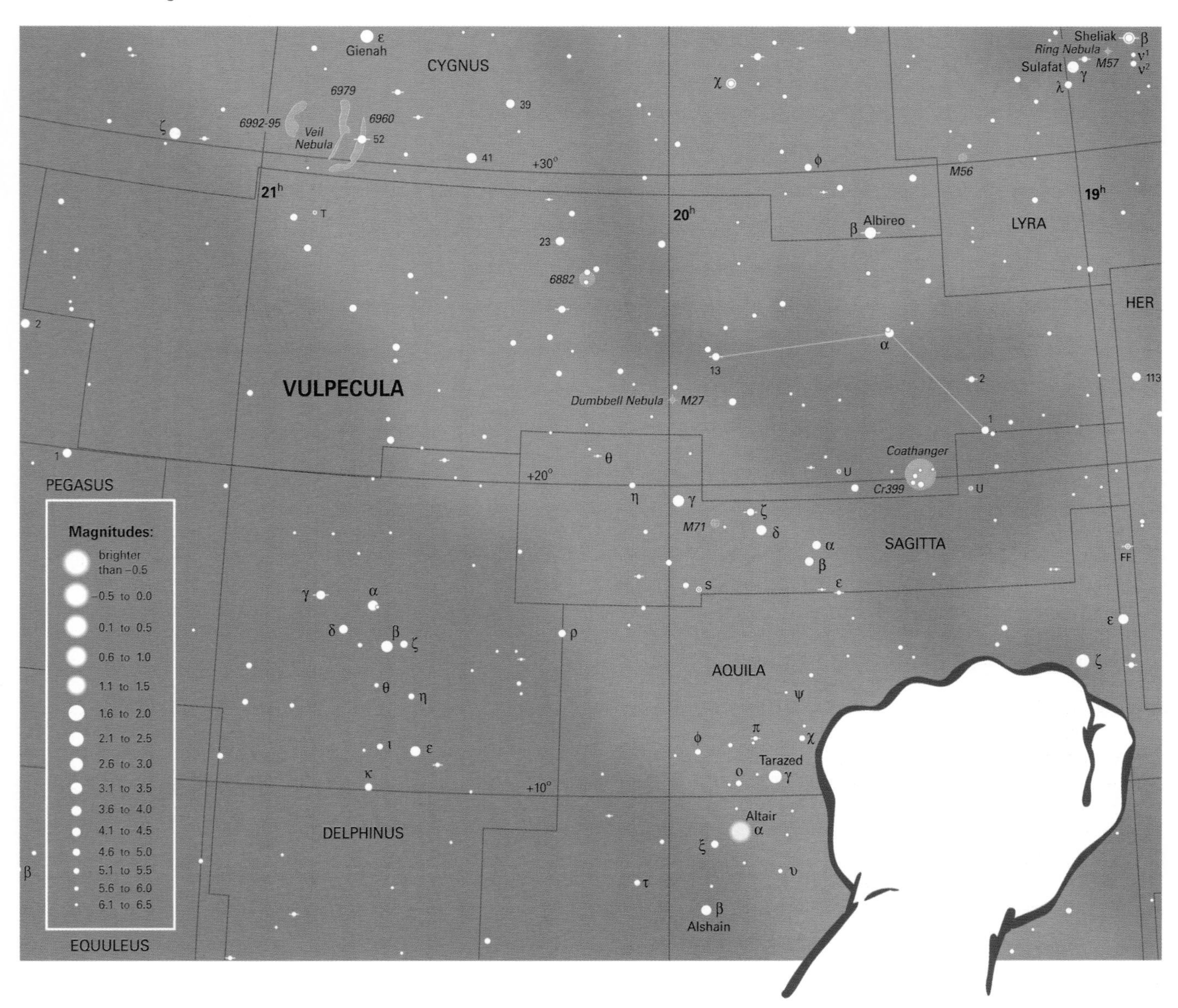

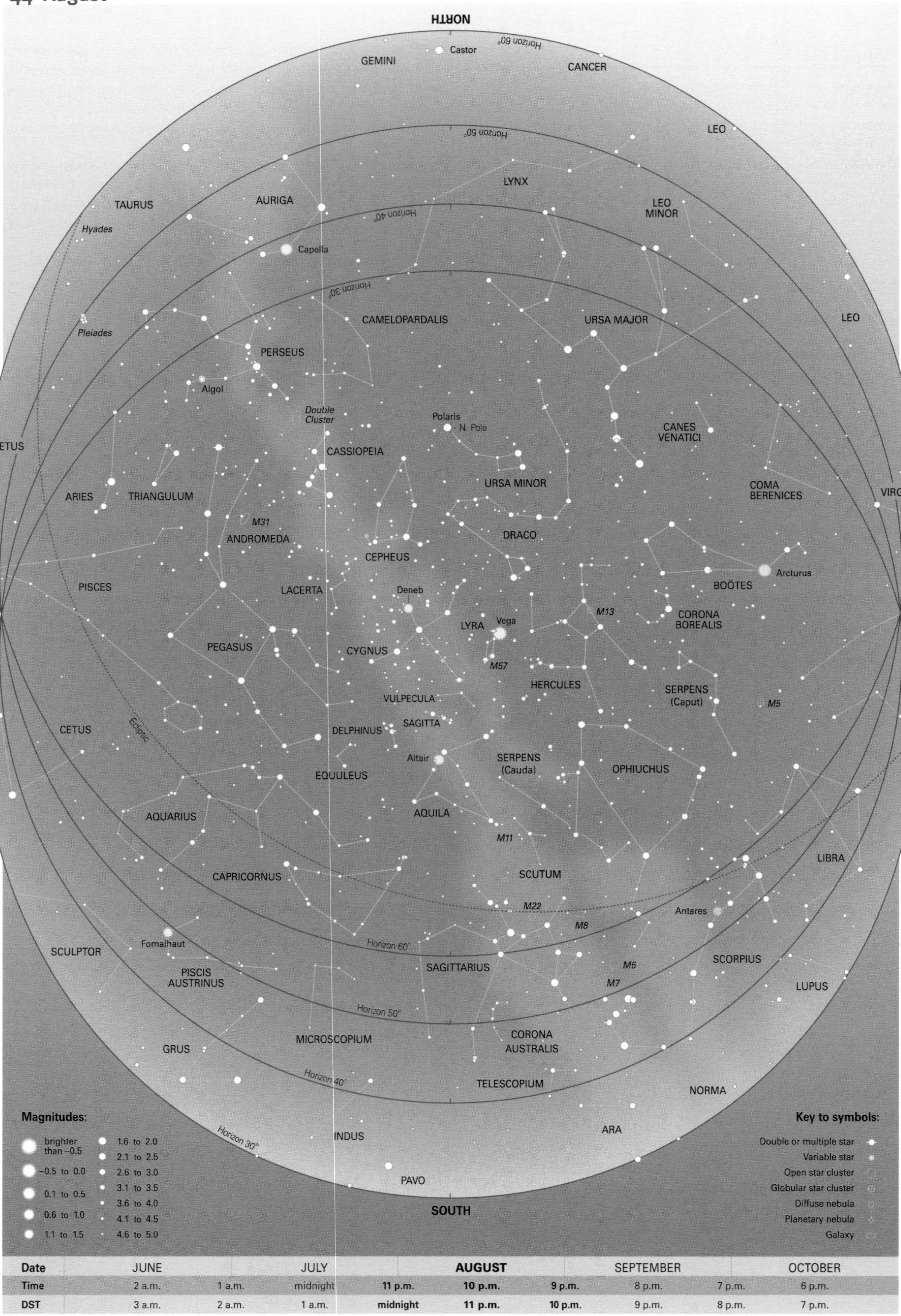

NORTH

Horizon 60°
Castor
GEMINI
CANCER
Horizon 50°
LEO
LYNX
TAURUS
AURIGA
Hyades
Horizon 40°
LEO MINOR
Capella
Horizon 30°
CAMELOPARDALIS
URSA MAJOR
LEO
Pleiades
PERSEUS
Algol
Double Cluster
Polaris
N. Pole
CANES VENATICI
CETUS
CASSIOPEIA
ARIES
TRIANGULUM
URSA MINOR
COMA BERENICES
VIRGO
M31
ANDROMEDA
DRACO
CEPHEUS
Arcturus
PISCES
LACERTA
Deneb
BOÖTES
EAST
M13
CORONA BOREALIS
WEST
LYRA
Vega
PEGASUS
CYGNUS
M57
HERCULES
SERPENS (Caput)
M5
VULPECULA
CETUS
Ecliptic
DELPHINUS
SAGITTA
Altair
SERPENS (Cauda)
EQUULEUS
OPHIUCHUS
AQUARIUS
AQUILA
M11
LIBRA
CAPRICORNUS
SCUTUM
M22
M8
Antares
Fomalhaut
Horizon 60°
SCULPTOR
SAGITTARIUS
SCORPIUS
M6
PISCIS AUSTRINUS
M7
LUPUS
Horizon 50°
GRUS
MICROSCOPIUM
CORONA AUSTRALIS
Horizon 40°
TELESCOPIUM
NORMA
Horizon 30°
INDUS
ARA
PAVO
SOUTH

Magnitudes:

brighter than −0.5
−0.5 to 0.0
0.1 to 0.5
0.6 to 1.0
1.1 to 1.5
1.6 to 2.0
2.1 to 2.5
2.6 to 3.0
3.1 to 3.5
3.6 to 4.0
4.1 to 4.5
4.6 to 5.0

Key to symbols:

Double or multiple star
Variable star
Open star cluster
Globular star cluster
Diffuse nebula
Planetary nebula
Galaxy

Date	JUNE		JULY		AUGUST		SEPTEMBER		OCTOBER
Time	2 a.m.	1 a.m.	midnight	11 p.m.	10 p.m.	9 p.m.	8 p.m.	7 p.m.	6 p.m.
DST	3 a.m.	2 a.m.	1 a.m.	midnight	11 p.m.	10 p.m.	9 p.m.	8 p.m.	7 p.m.

August

Key Stars

The Summer Triangle of Vega, Deneb and Altair stands high above as the sky darkens on August evenings. The ruddy glow of Arcturus is sinking towards the western horizon, and the stars of the Plough dip low in the north-west. Antares still glints in the south-west for those below latitude 50° north, while in the south-east lonely Fomalhaut climbs above the horizon in its barren part of the sky. In the east rises the Square of Pegasus, followed by the stars of Andromeda. Capella is above the northern horizon for observers north of latitude 40°.

The planets this month

Venus

2007 Too close to the Sun for observation. At inferior conjunction (between Earth and Sun) on August 18.
2008 Swamped by morning twilight throughout the month.
2009 Prominent in the eastern dawn sky at mag. −4.0.
2010 Prominent but low in the western evening twilight throughout the month at mag. −4.3. Passes below fainter Saturn in the first two weeks and below fainter Mars in the second and third weeks. At greatest elongation (maximum separation) from the Sun of 46° on August 20.
2011 At superior conjunction (behind the Sun) on August 16 and unobservable throughout the month.

Mars

2007 A morning object in Taurus, passing between the Pleiades and Hyades clusters in mid-month at mag. 0.4.
2008 Swamped by twilight in the western evening sky.
2009 A first-magnitude object in the eastern morning sky, passing from Taurus into Gemini.
2010 Lost in the western evening twilight for all but the most southerly latitudes.
2011 In the eastern dawn sky among the stars of Gemini at mag. 1.4.

Jupiter

2007 An evening object, stationary in southern Ophiuchus at mag. −2.3.
2008 Prominent for much of the night in Sagittarius at mag. −2.6.
2009 Visible all night in Capricornus. At opposition (due south at midnight) on August 14 at mag. −2.9, 603 million km (374 million miles) from Earth.
2010 Prominent for most of the night in Pisces at mag. −2.8.
2011 A morning object, stationary in Aries at mag. −2.5.

Saturn

2007 Too close to the Sun for observation. At conjunction (behind the Sun) on August 21.
2008 Lost in evening twilight throughout the month.
2009 Low in the western evening sky, becoming lost in twilight after mid-month.
2010 Starts the month low in the western evening twilight to the right of fainter Mars and above brilliant Venus, but becomes lost from view as it moves closer to the Sun.
2011 Low in the western evening twilight, becoming lost from view to more northerly observers by month's end.

Eclipses

Sun

2008 August 1. Total solar eclipse, visible from central and north Asia and the Arctic; maximum duration of totality 2m 27s. A partial eclipse is visible over Europe apart from the south and west, the Arctic, Asia except for the east and south-east, and the Middle East.
http://sunearth.gsfc.nasa.gov/eclipse/solar.html

Moon

2007 August 28. Total lunar eclipse, visible from North America except extreme east, western South America, Pacific Ocean, Australasia and east Asia. Totality starts 09.52 GMT, ends 11.22 GMT.
http://sunearth.gsfc.nasa.gov/eclipse/lunar.html

2008 August 16. Partial lunar eclipse, maximum 81%, visible from Europe, Asia except extreme north and east, Africa, Australia and eastern South America. Starts 19.36 GMT, ends 22.45 GMT.
http://sunearth.gsfc.nasa.gov/eclipse/lunar.html

August meteors

Warm summer nights are ideal for meteor observing, and August has the brightest of the year's meteor showers, the **Perseids**. This is the shower on which to begin meteor observing. The Perseids reach a peak around August 12 every year, when as many as one meteor a minute may be seen streaking away from the northern part of the constellation Perseus, near the border with Cassiopeia. The radiant does not rise very high until late at night, so do not expect to see many Perseids before midnight. Continue watching into the early morning hours, for the rate of Perseids should increase towards dawn as the radiant climbs higher in the sky. Perseid meteors can be spectacular. They are usually bright, they frequently flare up and many of them leave glowing trains. The shower is very broad, so that considerable numbers of Perseids can be seen for over a week before and after maximum. They are caused by dust from comet Swift–Tuttle, which has an orbital period of about 135 years; it last appeared in 1992.

Cygnus

Everyone has heard of the Southern Cross, as the southern constellation Crux is popularly called, but it is less well known that there is a far larger cross in the northern sky. The proper name for this constellation is Cygnus, the Swan, but it is often informally termed the Northern Cross.

The Greeks said that Cygnus represented the swan into which the god Zeus transformed himself in order to seduce Leda, wife of King Tyndareus of Sparta. On summer nights, Cygnus can be seen flying along the sparkling band of the Milky Way, its long neck outstretched towards the south-west and its stubby wings tipped by the stars Delta (δ) and Epsilon (ε) Cygni.

The swan's tail is marked by Alpha (α) Cygni, better known as **Deneb**, a name that comes from the Arabic for 'tail'. Deneb, of magnitude 1.3, is the faintest of the three stars that make up the **Summer Triangle**. But whereas the other two stars of the Triangle, Vega and Altair, are among the closest stars to the Sun, Deneb lies far off, over 3000 light years away. It is the most distant 1st-magnitude star, four times farther than Rigel in second place.

To be easily visible over such an immense distance, Deneb must be exceptionally powerful. It is in fact among the most luminous of stars, a supergiant with a white-hot surface giving out over 250,000 times as much light as the Sun. If Deneb were as close to us as the brightest star, Sirius, it would appear as brilliant as a half Moon, so the sky would never be dark when it was above the horizon.

Even more remarkable is the star **P Cygni**, located near the base of the swan's neck. It is currently of 5th magnitude, but on two occasions during the seventeenth century it temporarily brightened fivefold to 3rd magnitude. P Cygni is evidently so massive that it is unstable, throwing off shells of gas at irregular intervals, causing its surges in brightness. It is currently evolving into a red supergiant and will probably end its life by exploding as a supernova. Although data about such stars are highly uncertain, P Cygni is thought to have a mass of about 50 Suns and to shine as brightly as 100,000 Suns. It lies about twice as far away as Deneb, among the distant stars of the Milky Way.

Cygnus is an ideal region to explore on warm summer evenings. Begin by sweeping with binoculars along the star-splashed Milky Way, particularly rich in this area. Look for a major division in the Milky Way called the **Cygnus Rift**, caused by a lane of dark dust in the local spiral arm of our Galaxy that blocks starlight from behind. The rift can be traced southwards from Cygnus through Aquila and into Ophiuchus, broadening as it approaches us.

Of the star clusters in Cygnus, the best in binoculars is **M39**, a scattered group of 20 or so stars of 7th magnitude and fainter, arranged in a triangular shape. M39 lies about 900 light years away, much closer than the background stars of the Milky Way.

Under clear, dark skies binoculars should show the large, misty shape of the **North America Nebula**, so-called because of its remarkable resemblance to the continent of North America, although its shape is more apparent on long-exposure photographs than through small instruments. The Nebula, also known as **NGC 7000**, lies near the star Xi (ξ) Cygni, and covers three times the width of the Moon, so large that only binoculars or wide-angle telescopes can encompass it all.

This immense mass of gas and dust, nearly 50 light years in diameter, appears to lie near the star Deneb. It has often been suggested that the North America Nebula is lit up by Deneb, but in fact the Nebula seems to be only about half the distance of Deneb, so that other stars hidden within the Nebula must instead be responsible for making it glow.

While scanning Cygnus, be sure to look at **Omicron-1** (o^1) **Cygni**, a binocular gem. This orange star, magnitude 3.8, forms a beautiful wide pairing with the turquoise 30 Cygni, one magnitude fainter. Binoculars, if held steadily, will also reveal a 7th-magnitude blue star closer to Omicron-1 Cygni.

This is an ideal warm-up before turning your gaze upon the queen of double stars, the beautiful **Beta** (β) **Cygni**, or **Albireo**. (Its name, incidentally, results from a mis-translation and is completely meaningless.) Albireo is easily found, marking the beak of the swan – or, if you prefer, the foot of the cross. The smallest of telescopes, and even binoculars mounted steadily, show that Albireo consists of a glorious pair of amber and blue-green stars, magnitudes 3.1 and 5.1, like a celestial traffic light. These beautifully contrasting colours, similar to those of Omicron-1 Cygni, will appear more intense if you put the stars slightly out of focus, or if you tap the telescope so that the image vibrates.

Although the two components of Albireo lie at similar distance from us, about 380 light years, it is not certain that they form a true binary. If the fainter star is in orbit around the brighter one, it would probably take over 100,000 years to complete one circuit. Whatever the case, Albireo remains one of the greatest attractions on any Grand Tour of the sky.

For another celebrated double, look above the crossbar of Cygnus to find **61 Cygni**, a pair of orange stars of 5th and 6th magnitude, easily separated in small telescopes. The two stars orbit each other every 650 years. Considerable historical interest surrounds 61 Cygni, for it was the first star to have its distance measured by the technique of parallax. In this technique, a star's position is accurately noted on two occasions six months apart, when the Earth is on opposite sides of its orbit around the Sun. The slight shift in the star's position as seen from the two vantage points reveals how far away the star is, by simple geometry. A nearby star will move noticeably, whereas a distant star will hardly move at all.

The observations of 61 Cygni were made by the German astronomer Friedrich Wilhelm Bessel in 1838. He concluded that 61 Cygni is 10.3 light years away, a figure that compares well with the modern value of 11.4 light years. Of the naked-eye stars, only Alpha Centauri, Sirius and Epsilon Eridani are closer than 61 Cygni.

A less well-known treasure of Cygnus is a remarkable planetary nebula, **NGC 6826**, popularly known as the **Blinking Planetary** because it appears to blink on and off as you look at it and then away from it. To find it, first locate **16 Cygni**, an attractive duo of 6th-magnitude yellow stars easily split by small telescopes. The Blinking Planetary lies less than 1° from this pair, so that both objects will just fit into the same telescopic field of view.

In small telescopes with low magnification NGC 6826 looks like a fuzzy, faint star. Higher magnification reveals a light-blue disk with a 10th-magnitude star at its centre, although this central star will probably be beyond the reach of the smallest telescopes. Like any delicate object, the nebula is best seen with averted vision – that is, by looking to one side of it. Alternately looking at the nebula and away again produces the uncanny blinking effect that gives this object its popular name. Needless to say, the blinking is purely an optical effect in the observer's eye and does not originate in the nebula itself.

Some of the most amazing objects in Cygnus are, unfortunately, beyond the reach of amateur instruments. On the border of Cygnus with Vulpecula lies an enormous loop of gas, the **Veil Nebula**, the twisted wreckage of a star that exploded as a supernova some 5000 years ago. At an estimated distance of 1400 light years from us, the supernova would have been bright enough to cast shadows on Earth. The Veil Nebula covers 3° of sky, equivalent to the width of two fingers held at arm's length. The brightest portion of the Veil, **NGC 6992–95**, may just be made out in binoculars under exceptional conditions, but most observers will have to content themselves with studying photographs of it in books.

Between Gamma (γ) and Delta (δ) Cygni, in one of the wings of the Swan, lies **Cygnus A**, one of the most intense sources of radio noise in the sky. Photographs show that it is a peculiar-looking elliptical galaxy of 15th magnitude, either exploding or undergoing a collision. It lies deep in the Universe, about 750 million light years outside the Milky Way.

Of all the objects in Cygnus the most bizarre is **Cygnus X-1**, one of the best candidates in the Milky Way for a black hole. Although the black hole itself is invisible, it gives itself away by sucking in hot gas from a neighbouring 9th-magnitude star, HDE 226868. Gas falling towards the black hole heats up to millions of degrees, emitting X-rays that are detected by satellites. Cygnus X-1 lies about 8000 light years away, along the same spiral arm of the Galaxy in which the Sun lies. Its position is about half a degree (one Moon's breadth) from Eta (η) Cygni. Not far away, midway between Eta (η) and Epsilon (ε) Cygni, lies another black hole suspect, a variable X-ray source called V404 Cygni. In this case, the black hole is orbited by a faint orange star, gas from which falls towards the black hole.

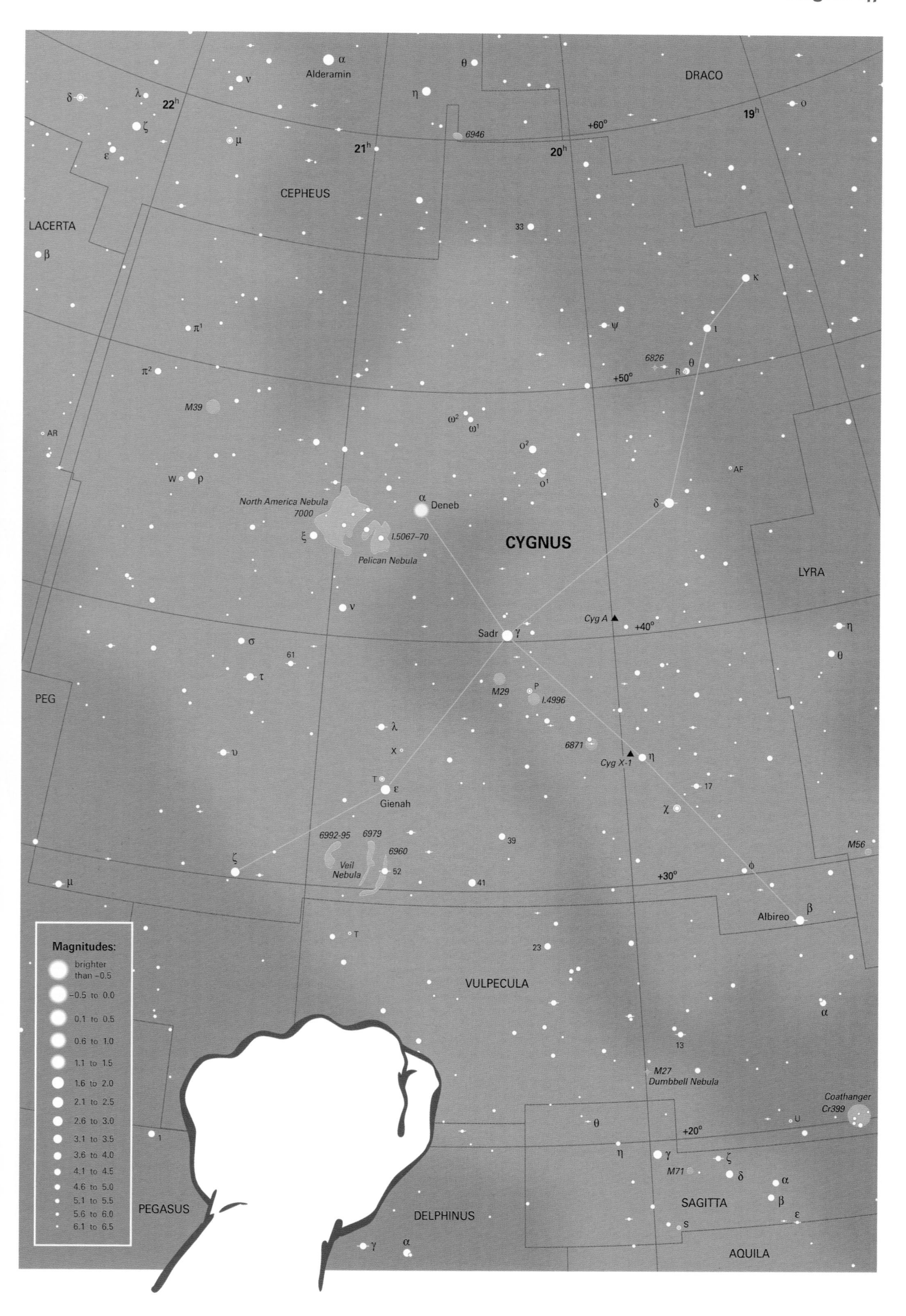

DRACO
CEPHEUS
LACERTA
Alderamin
CYGNUS
LYRA
North America Nebula
7000
Pelican Nebula
Deneb
Sadr
Cyg A
Cyg X-1
Gienah
Veil
Nebula
Albireo
PEG
VULPECULA
M27
Dumbbell Nebula
Coathanger
Cr399
SAGITTA
PEGASUS
DELPHINUS
AQUILA
Magnitudes:
brighter than −0.5
−0.5 to 0.0
0.1 to 0.5
0.6 to 1.0
1.1 to 1.5
1.6 to 2.0
2.1 to 2.5
2.6 to 3.0
3.1 to 3.5
3.6 to 4.0
4.1 to 4.5
4.6 to 5.0
5.1 to 5.5
5.6 to 6.0
6.1 to 6.5

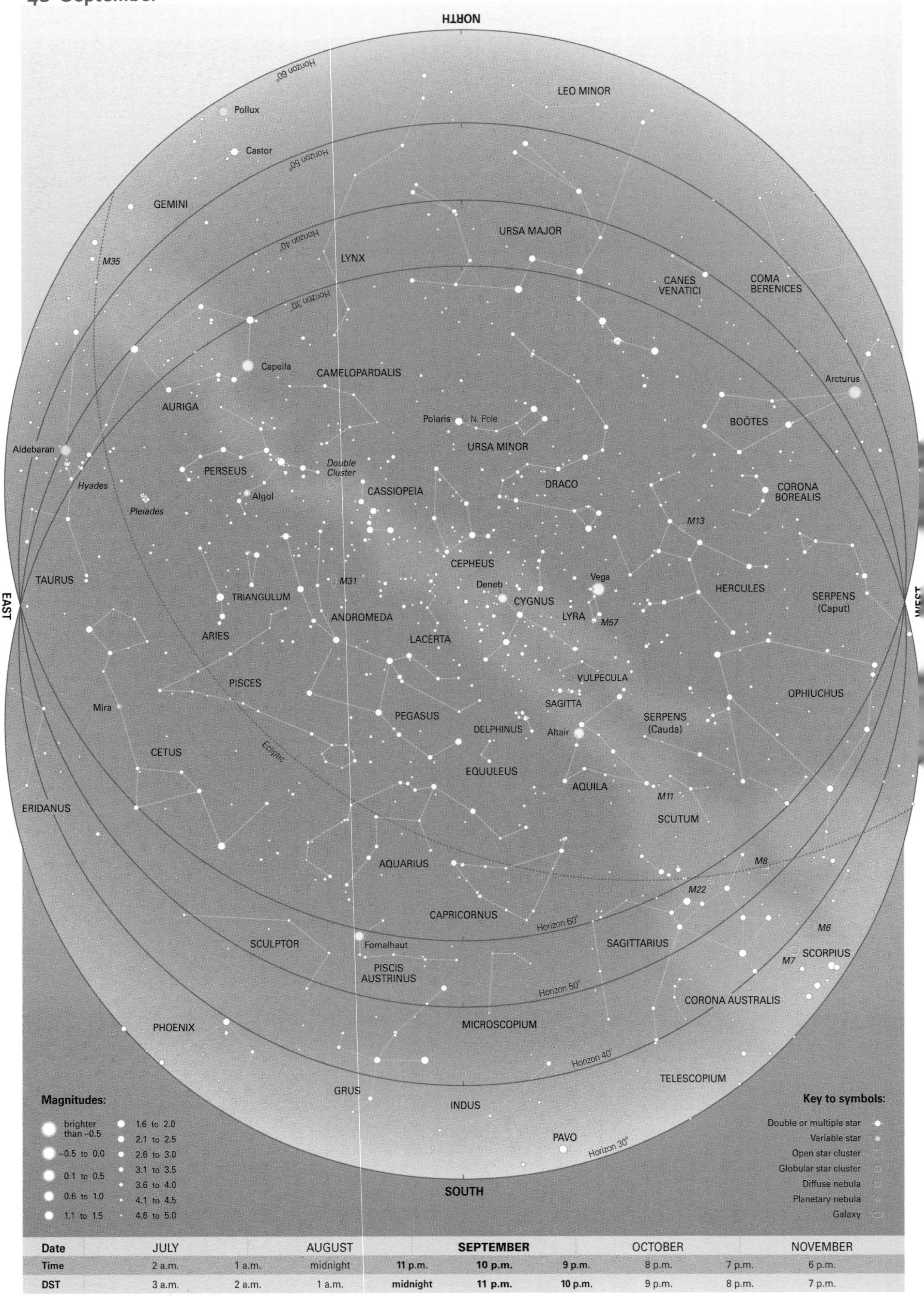

NORTH
EAST
WEST
SOUTH

Horizon 60°
Horizon 50°
Horizon 40°
Horizon 30°

LEO MINOR
Pollux
Castor
GEMINI
M35
LYNX
URSA MAJOR
CANES VENATICI
COMA BERENICES
Capella
CAMELOPARDALIS
AURIGA
Arcturus
Polaris
N. Pole
URSA MINOR
BOÖTES
Aldebaran
PERSEUS
Double Cluster
Hyades
Algol
CASSIOPEIA
DRACO
CORONA BOREALIS
Pleiades
M13
CEPHEUS
TAURUS
M31
Deneb
Vega
HERCULES
SERPENS (Caput)
TRIANGULUM
CYGNUS
ANDROMEDA
LYRA
M57
ARIES
LACERTA
PISCES
VULPECULA
OPHIUCHUS
SAGITTA
Mira
PEGASUS
SERPENS (Cauda)
DELPHINUS
Altair
CETUS
Ecliptic
EQUULEUS
AQUILA
M11
ERIDANUS
SCUTUM
AQUARIUS
M8
M22
CAPRICORNUS
M6
SCULPTOR
Fomalhaut
SAGITTARIUS
SCORPIUS
M7
PISCIS AUSTRINUS
CORONA AUSTRALIS
PHOENIX
MICROSCOPIUM
TELESCOPIUM
GRUS
INDUS
PAVO

Magnitudes:

- brighter than −0.5
- −0.5 to 0.0
- 0.1 to 0.5
- 0.6 to 1.0
- 1.1 to 1.5
- 1.6 to 2.0
- 2.1 to 2.5
- 2.6 to 3.0
- 3.1 to 3.5
- 3.6 to 4.0
- 4.1 to 4.5
- 4.6 to 5.0

Key to symbols:

- Double or multiple star
- Variable star
- Open star cluster
- Globular star cluster
- Diffuse nebula
- Planetary nebula
- Galaxy

Date	JULY		AUGUST		**SEPTEMBER**		OCTOBER		NOVEMBER
Time	2 a.m.	1 a.m.	midnight	**11 p.m.**	**10 p.m.**	**9 p.m.**	8 p.m.	7 p.m.	6 p.m.
DST	3 a.m.	2 a.m.	1 a.m.	**midnight**	**11 p.m.**	**10 p.m.**	9 p.m.	8 p.m.	7 p.m.

September

Key Stars

The Summer Triangle remains prominent as darkness falls in September skies. Arcturus is vanishing below the north-western horizon. From the Summer Triangle the stardust trail of the Milky Way leads eastwards to Cassiopeia and Perseus, while Capella twinkles brightly in the north-east. The Square of Pegasus is high in the south-east, while below it lies lonely Fomalhaut, just brushing the southern horizon for far-northern observers. In the east Aldebaran and the stars of Taurus appear, heralding the autumn. In the north, the Plough sinks low.

The planets this month

Venus

2007 Emerges into the eastern dawn sky at the start of the month and rapidly becomes a prominent morning object at mag. −4.5.
2008 Low in the western evening sky throughout the month at mag. −3.9.
2009 In the eastern dawn sky all month at mag. −3.9.
2010 Low in the south-west evening sky throughout the month at mag. −4.5, becoming lost in twilight from more northerly latitudes.
2011 Too close to the Sun to be seen this month.

Mars

2007 In Taurus, ending the month on the border with Gemini at mag. 0.0.
2008 Lost in evening twilight, low in the west.
2009 A morning object in Gemini at mag. 0.9.
2010 Lost in evening twilight, low in the west.
2011 In the eastern morning sky, moving from Gemini into Cancer at mag. 1.4. Ends the month near the Beehive cluster.

Jupiter

2007 In the south-west evening sky at mag. −2.1.
2008 A prominent evening object, stationary in Sagittarius at mag. −2.4.
2009 Prominent for most of the night in Capricornus at mag. −2.8.
2010 Visible all night in Pisces. Opposition (due south at midnight) is on September 21 at mag. −2.9, 592 million km (368 million miles) from Earth.
2011 Visible for most of the night, stationary in Aries at a prominent mag. −2.8.

Saturn

2007 Emerges into the eastern dawn sky after mid-month at mag. 0.7.
2008 At conjunction (behind the Sun) on September 4, but emerges rapidly into the eastern dawn sky in the second half of the month at mag 0.9.
2009 Too close to the Sun for observation all month. At conjunction (behind the Sun) on September 17.
2010 Too close to the Sun for observation.
2011 Lost in evening twilight throughout the month.

Eclipses

Sun

2007 September 11. Partial solar eclipse, maximum 74%, visible from southern South America, south-western Atlantic Ocean, part of Antarctica.
http://sunearth.gsfc.nasa.gov/eclipse/solar.html

Cassiopeia

The only married couple among the constellations are Cepheus and Cassiopeia. In Greek mythology they were King and Queen of Ethiopia, parents of Andromeda. According to fable, Cassiopeia once boasted that she was more beautiful than the group of sea nymphs called the Nereids. This incurred the wrath of the sea god Poseidon, who sent a terrible monster to ravage the lands of King Cepheus. To save his country, Cepheus was forced to offer his daughter Andromeda as a sacrifice to the monster, although she was rescued from the monster's jaws by Perseus.

Cassiopeia and Cepheus are to be found next to each other in the north polar region of the sky. Cassiopeia, the more prominent of the husband-and-wife pair, is depicted sitting in a chair. She circles incongruously around the pole, appearing to hang upside down for part of the night, a lesson in humility decreed by the Nereids.

Cassiopeia is one of the easiest constellations to recognize because of its distinctive W-shape. The constellation is easily located, opposite the celestial pole from the Plough. **Gamma** (γ) **Cassiopeiae**, the central star of the W-shape, is worth keeping an eye on for its unpredictable brightness, which has fluctuated between magnitudes 1.6 and 3.0 since early last century, when its variability was first noted. Compare it with neighbouring Alpha (α) Cas, magnitude 2.2, and Delta (δ) Cas, magnitude 2.7. Gamma Cassiopeiae is a hot, blue star rotating so rapidly that it throws off shells of gas, thus producing the changes in brightness. As with all such peculiar stars data are uncertain, but Gamma Cas is thought to give out as much light as 5000 Suns and to lie about 600 light years away.

A beautiful double star for small telescopes is **Eta** (η) **Cassiopeiae**, consisting of yellow and reddish stars, magnitudes 3.5 and 7.5. These two stars orbit each other every 480 years. As seen from Earth they were closest together in 1889 and will continue to move apart until the middle of next century, making the double progressively easier to divide. The pair are only 19 light years from us, relatively close on the interstellar scale.

Lying in a rich part of the Milky Way, Cassiopeia contains numerous star clusters. Best is **M52**, easily found near the border with Cepheus next to a 5th-magnitude star, 4 Cassiopeiae. M52 is a fuzzy patch in binoculars that splits into a field of faint stars when seen through a telescope. One member, an 8th-magnitude orange giant, is brighter than the rest. Another cluster visible as a fuzzy patch in binoculars is **M103**. Telescopes show it to have an elongated shape with a spine of stars. The brightest of these, a double star at the northern end of the spine, is not in fact a true member of the cluster, but actually lies in the foreground. Nearby is **NGC 663**, a scattered group of stars of various brightnesses, larger and more prominent in binoculars than M103.

Look also at the cluster **NGC 457**, next to the 5th-magnitude star Phi (ϕ) Cassiopeiae, which is apparently a true member of the cluster. Since the cluster lies nearly 10,000 light years away, Phi

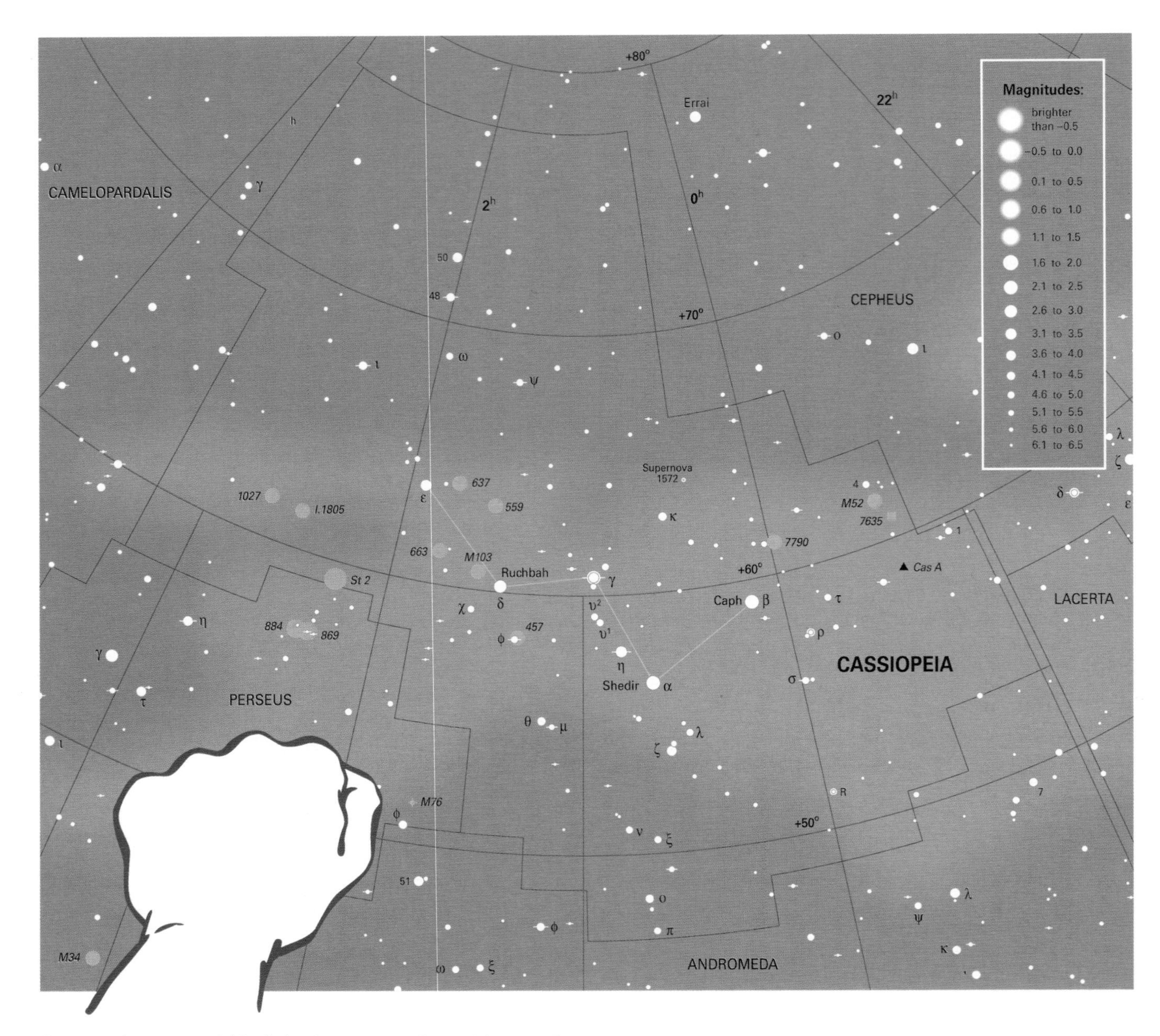

Cas must be an astonishingly luminous supergiant, giving out the light of over 250,000 Suns, five times that of Rigel. As seen through a telescope, the stars of NGC 457 appear to be arranged in chains.

Even more remarkable than Phi is **Rho** (ρ) **Cassiopeiae**. At first glance it appears unexceptional, but it turns out to be one of the most luminous stars known – a yellow-white supergiant shining as brightly as half a million Suns and lying some 10,000 light years away, perhaps more. It varies in brightness between 4th and 6th magnitude every 11 months or so as it pulsates in size.

Two other objects in Cassiopeia deserve mention, although both are beyond the reach of amateur instruments. They are the remains of exploded stars, or supernovae. One, known as **Tycho's Star**, flared up near Kappa (κ) Cassiopeiae in 1572. At its peak it became as bright as Venus, and remained visible to the naked eye for 16 months. It was carefully observed by the great Danish astronomer Tycho Brahe, after whom it is named. All that is left of Tycho's Star are some faint wisps of gas, visible only with large telescopes, coinciding with a source of radio noise.

The strongest radio source in the sky, **Cassiopeia A**, lies almost 3° south of the cluster M52. Cassiopeia A comprises the remains of another supernova, estimated to have exploded in the late seventeenth century. But apparently it was abnormally faint and did not become visible to the naked eye. Why it was so faint is not known.

Cepheus

Cepheus, representing the husband of Cassiopeia, lies close to the north celestial pole. In shape it resembles a squat tower with a steeple. Look first beneath the base of the tower to find **Mu** (μ) **Cephei**, one of the most strongly coloured stars in the sky. Sir William Herschel called it the **Garnet Star** because of its deep red tint, which is notable in binoculars and small telescopes. Mu Cephei is a red supergiant with the luminosity of 50,000 Suns. Like many red supergiants and giants it is erratically variable, fluctuating between magnitudes 3.4 and 5.1. Take a look at this star next time you are out, and try to estimate its brightness by comparing it with Zeta (ζ) Cephei, magnitude 3.4, Epsilon (ε) Cephei, magnitude 4.2, and Lambda (λ) Cephei, magnitude 5.0.

In the middle of the tower of Cepheus lies **Xi** (ξ) **Cephei**, a tidy double star for small telescopes. Its components are of magnitudes 4.4 and 6.5, colours white and yellow. The two stars form a genuine binary, orbiting each other every 4000 years or so. Next try **Beta** (β) **Cephei**, a magnitude 3.2 star with an 8th-magnitude companion, difficult to distinguish in the smallest telescopes because of the brightness difference. Beta itself is slightly variable in brightness, by about 0.1 magnitude, virtually undetectable to the naked eye.

Finally we come to the celebrity of the constellation, **Delta** (δ) **Cephei**. A small telescope shows it as an easy and attractive double, consisting of a 4th-magnitude yellow primary whose magnitude 6.3 companion, bluish in colour, is wide enough to be picked out in binoculars, if held steadily. But the greatest interest of this star is the primary, a yellow supergiant that varies regularly every 5 days 9 hours. At its brightest, Delta Cephei reaches magnitude 3.5, twice as bright as when at its faintest, magnitude 4.4. These variations can easily be tracked with the naked eye. They are caused by pulsations in the size of the star, like a slowly beating heart. In the case of Delta Cephei, its diameter changes from 40 to 45 times that of the Sun.

The variability of Delta Cephei was discovered in 1784 by John Goodricke, a 19-year-old amateur astronomer in York, England. Tragically, while continuing his observations of this star, Goodricke caught pneumonia and died, age 21.

Delta Cephei is the prototype of a class of variable stars known simply as **Cepheids**, which are of particular importance to astronomers. Cepheids are supergiants with surface temperatures of 5000 K to 6000 K, similar to that of the Sun, so they appear yellowish in colour. But being much bigger than the Sun, they are also very much brighter (5000 times brighter than the Sun in the case of Delta Cephei) and hence they are easily visible over great distances. Delta Cephei itself is about 1000 light years from us.

Cepheids vary in brightness by up to two magnitudes (i.e. up to 6.3 times) with periods ranging from about 1 to 135 days. Over 700 Cepheids are now known in our Galaxy and they have been discovered in other galaxies, too. The importance of Cepheids to astronomers is that there is a direct link between their inherent brightness (their 'wattage') and their period of variability. The brighter the Cepheid, the longer it takes to vary. By using this relationship, astronomers can deduce the wattage of a given Cepheid simply by monitoring the period of its light changes.

But a star's brightness as seen from Earth is affected by its distance from us. So astronomers can work out how far away a Cepheid is by comparing the star's computed wattage with the brightness at which it appears in the sky. Hence Cepheids serve as marker beacons for measuring distances in space.

Suitable comparison stars for following the light changes of Delta Cephei are Zeta (ζ), Epsilon (ε) and Lambda (λ) Cephei, the same trio as used for Mu Cephei. These stars are shown on the chart below.

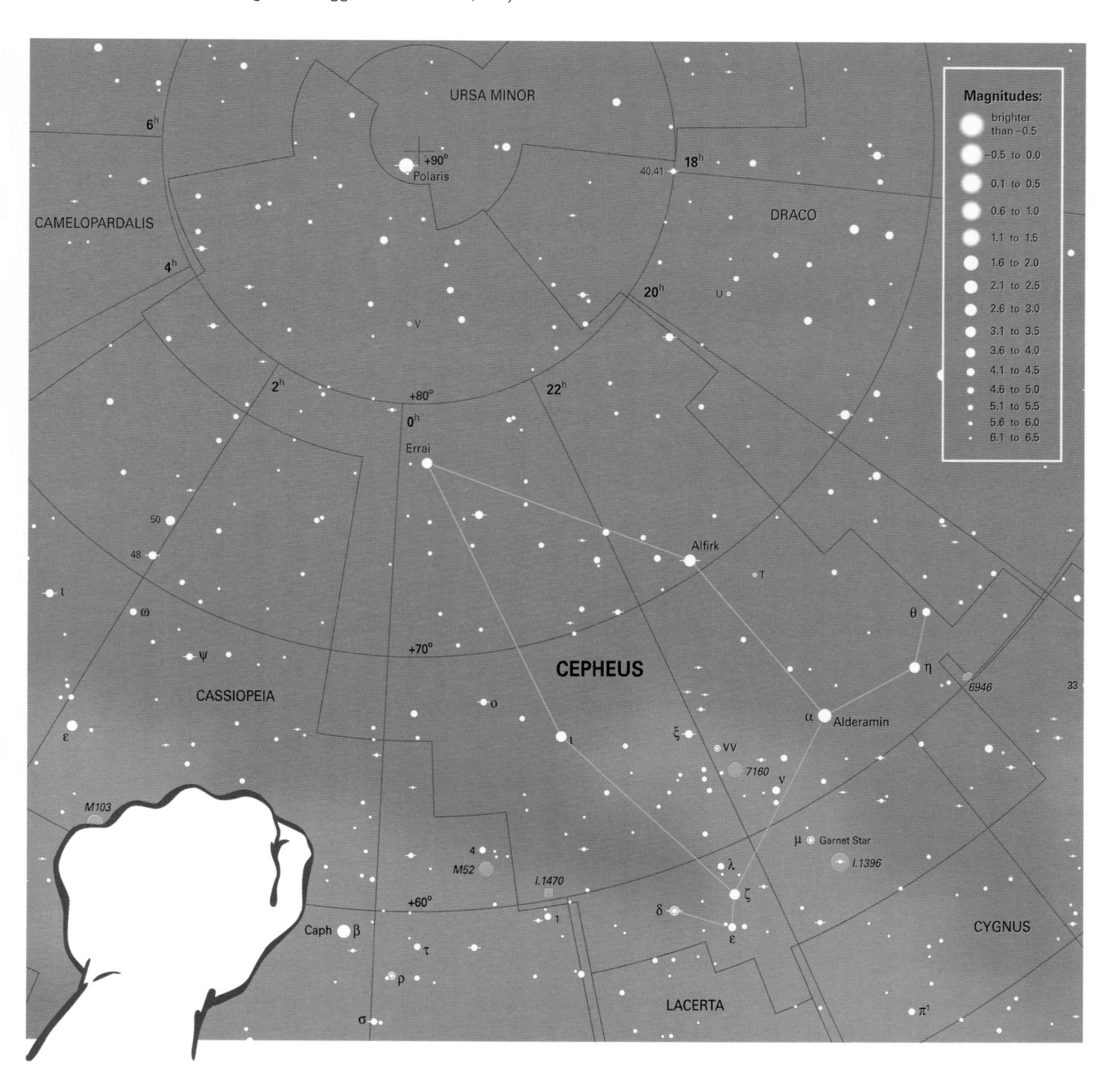

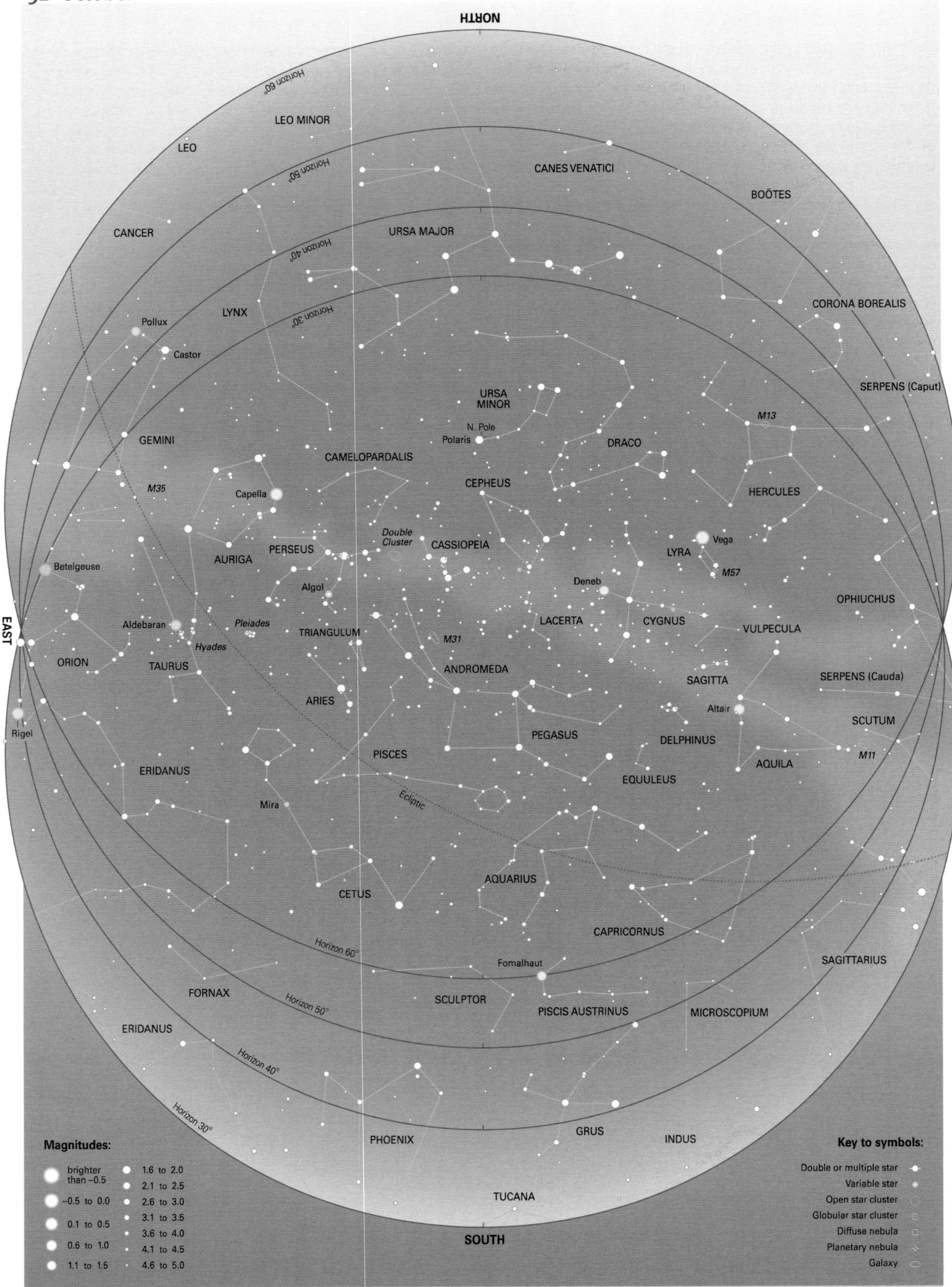

NORTH

Horizon 60°
Horizon 50°
Horizon 40°
Horizon 30°

LEO MINOR
LEO
CANES VENATICI
BOÖTES
CANCER
URSA MAJOR
CORONA BOREALIS
LYNX
Pollux
Castor
SERPENS (Caput)
URSA MINOR
N. Pole
Polaris
M13
GEMINI
DRACO
CAMELOPARDALIS
M35
CEPHEUS
Capella
HERCULES
Double Cluster
CASSIOPEIA
Vega
PERSEUS
LYRA
AURIGA
M57
Betelgeuse
Algol
Deneb
OPHIUCHUS
EAST
Aldebaran
Pleiades
LACERTA
CYGNUS
WEST
TRIANGULUM
VULPECULA
Hyades
M31
ORION
TAURUS
ANDROMEDA
SERPENS (Cauda)
SAGITTA
ARIES
Altair
Rigel
SCUTUM
PEGASUS
DELPHINUS
PISCES
M11
ERIDANUS
AQUILA
EQUULEUS
Mira
Ecliptic
AQUARIUS
CETUS
CAPRICORNUS
Horizon 60°
SAGITTARIUS
Fomalhaut
FORNAX
Horizon 50°
SCULPTOR
PISCIS AUSTRINUS
MICROSCOPIUM
ERIDANUS
Horizon 40°
Horizon 30°
PHOENIX
GRUS
INDUS
TUCANA
SOUTH

Magnitudes:

brighter than −0.5
−0.5 to 0.0
0.1 to 0.5
0.6 to 1.0
1.1 to 1.5
1.6 to 2.0
2.1 to 2.5
2.6 to 3.0
3.1 to 3.5
3.6 to 4.0
4.1 to 4.5
4.6 to 5.0

Key to symbols:

Double or multiple star
Variable star
Open star cluster
Globular star cluster
Diffuse nebula
Planetary nebula
Galaxy

Date	AUGUSTUS		SEPTEMBER	OCTOBER		NOVEMBER		DECEMBER
Time	2 a.m.	1 a.m.	midnight	**11 p.m.**	**9 p.m.**	8 p.m.	7 p.m.	6 p.m.
DST	3 a.m.	2 a.m.	1 a.m.	**midnight**	**10 p.m.**	9 p.m.	8 p.m.	7 p.m.

October

Key Stars

The Summer Triangle still lingers in the western sky, reluctant to give way to the stars of autumn. Due south, the Square of Pegasus stands high while Fomalhaut twinkles over the southern horizon. Between Pegasus and the north celestial pole, the W shape of Cassiopeia is well presented. To the east are Perseus and Auriga, embedded in the Milky Way's starry path. On the eastern horizon, Orion and Gemini are rising. The Plough dips low in the north, and is below the horizon for observers at latitude 30°.

The planets this month

Venus

2007 Prominent in the eastern morning sky among the stars of Leo at mag. –4.5. Passes fainter Saturn in mid-month. Greatest elongation (maximum separation) from the Sun of 46° is on October 28.

2008 Low in the south-west evening sky at mag. –4.0.

2009 In the eastern dawn sky at mag. –3.9, passing fainter Saturn in mid-month.

2010 Too close to the Sun for observation. At inferior conjunction (between Earth and Sun) on October 29.

2011 Emerges into the south-west evening sky by month's end, as seen from more southerly latitudes.

Mars

2007 In Gemini, brightening to mag. –0.6 at month's end.

2008 Too close to the Sun for observation all month.

2009 In the eastern morning sky, passing from Gemini into Cancer and brightening to mag. 0.5. Ends the month in front of the Beehive cluster.

2010 Too low in the south-western evening twilight to be seen from the most northerly latitudes.

2011 A morning object, starting the month near the Beehive cluster in Cancer and then moving into Leo at mag. 1.2.

Jupiter

2007 Low in the south-west evening sky at mag. –1.9.

2008 An evening object in Sagittarius at mag. –2.2.

2009 A prominent evening object at mag. –2.6, stationary in Capricornus.

2010 Prominent for most of the night on the Pisces–Aquarius border at mag. –2.8.

2011 Visible all night in southern Aries. At opposition (due south at midnight) on October 29 at mag. –2.9, 594 million km (369 million miles) from Earth.

Saturn

2007 In the eastern morning sky in Leo at mag. 0.8. Brilliant Venus passes around mid-month.

2008 In the eastern dawn sky at mag. 1.0.

2009 Emerges into the eastern dawn sky in the second week of October at mag. 1.1, passing to the left of brilliant Venus around mid-month.

2010 At conjunction (behind the Sun) on October 1. Emerges into the eastern dawn sky by month's end.

2011 Too close to the Sun for observation. At conjunction (behind the Sun) on October 13.

October meteors

One of the year's less spectacular meteor showers, the **Orionids**, radiates from near the Orion–Gemini border every October, reaching a peak of about 25 meteors per hour around October 21, although some activity can be seen for up to a week either side of this date. Orionid meteors are fast-moving and can leave dusty trains, but they are not particularly bright so a dark site will be needed to see them well. Like the Eta Aquarids in May, they are dust released from the famous Halley's Comet. The Orionid radiant does not rise until late, so observations are best made after midnight. The radiant lies due south around 4.30 a.m.

Andromeda

The most famous of all Greek myths, the story of Perseus and Andromeda, is depicted in the skies of autumn. It is the original version of George and the dragon. The tale tells how beautiful Andromeda was chained to a rock by her father, Cepheus, as a sacrifice to appease a terrible monster from the sea that was devastating his country's coastline. Fortunately the hero Perseus happened along in the nick of time, slew the monster and married Andromeda.

All the characters in this story – Andromeda, her parents Cassiopeia and Cepheus, and her rescuer Perseus – are represented by adjacent constellations. Even the sea monster lies nearby, in the form of the constellation Cetus.

Andromeda itself is not a particularly distinctive constellation. Its most noticeable feature is a curving line of stars that branches off from the top left corner of the Square of Pegasus, forming an immense dipper shape. In fact, the star at the top left corner of the Square of Pegasus is Alpha (α) Andromedae. In Greek times this star was considered common to both Andromeda and Pegasus, but is now assigned exclusively to Andromeda. Its two alternative names, Alpheratz and Sirrah, come from the Arabic meaning 'horse's navel', reflecting its former association with Pegasus.

Andromeda's body can be traced along the line of 2nd-magnitude stars from Alpha, which marks her head, via Beta (β), her waist, to Gamma (γ), her left foot chained to the rock. **Gamma Andromedae** is well worth a closer look, for it is an impressive double star with a fine colour contrast, like a tighter version of Albireo in Cygnus. The main star of Gamma Andromedae, magnitude 2.2, is an orange giant very similar to Arcturus. Small telescopes show that it has a 5th-magnitude blue companion. This companion star is itself a close double, but large telescopes are needed to separate the two components.

South of Gamma Andromedae is an even easier double, **56 Andromedae**, whose unrelated 6th-magnitude components are so wide apart they are divisible in binoculars. This pair of orange giant suns seemingly lies on the edge of **NGC 752**, a widespread cluster of faint stars, but in fact they are foreground objects, not members of the cluster.

Completing a triangle with Gamma and 56 Andromedae is **Upsilon (υ) Andromedae**, a 4th-magnitude star similar to the Sun which is worth locating because it is the first star around which a multi-planet system has been discovered. Three planets are known, with masses ranging from that of Saturn to nearly five

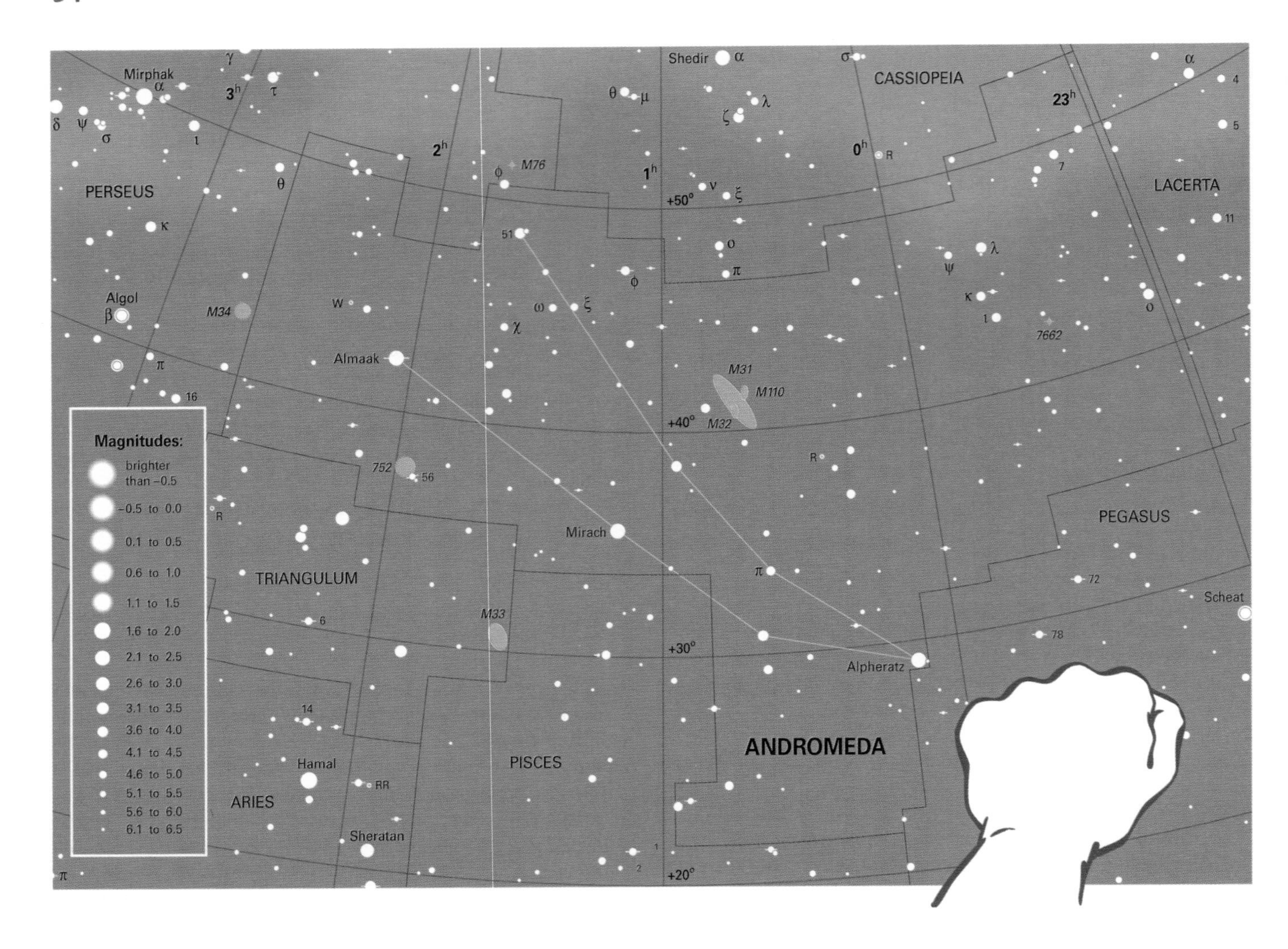

times that of Jupiter. Other planets of Earth-like mass may also exist but are so far undetected. Upsilon Andromedae is 44 light years away.

Andromeda contains one of the easiest planetary nebulae for small telescopes, **NGC 7662**. Look for it one Moon's breadth from the 6th-magnitude star 13 Andromedae. At first glance it resembles a fuzzy star of 9th magnitude. But higher powers show its slightly elliptical outline, similar in size to the disk of Mars, with the luminous blue-green tinge typical of planetary nebulae. Anyone who thinks that planetary nebulae require large apertures should try NGC 7662.

Overshadowing all else in Andromeda is the great spiral galaxy **M31**, a near-twin of our own Milky Way. A line from Beta (β) through Mu (μ) Andromedae points to it. Under even averagely clear skies, M31 appears to the naked eye as a faint elliptical glow, high overhead. M31 is the farthest object visible to the naked eye, some 2.5 million light years away. As you gaze upwards at M31, reflect that the light from it entering your eyes tonight has been travelling for two and a half million years, since the time when our ape-like ancestors were roaming the plains of Africa.

M31 has a special place in the history of astronomy, for it was the first object to be identified as a separate galaxy outside our own Milky Way. In 1924 the American astronomer Edwin Hubble used the 2.5-metre (100-inch) Mount Wilson reflector, then the largest telescope in the world, to take long-exposure photographs of M31. These photographs showed that it contained individual stars, but so faint that they must lie far beyond the stars of our Milky Way. Until then, M31 and objects like it had been generally assumed to be spiral-shaped clouds of gas within our own Milky Way. Now we know that the Milky Way is but one galaxy in a Universe full of countless other galaxies, stretching as far as the largest telescopes can see.

It is easy to understand why previous astronomers failed to recognize the true nature of M31. Even in large telescopes it appears as a nebulous patch with no hint of individual stars. Binoculars show it unmistakably, extending up to several Moon diameters, half the width of your outstretched fist. It appears elliptical in shape because we are seeing it at an angle. See how far you can trace its faint outer regions, for its extent will depend on sky conditions and the instrument used. Small telescopes tend to show only the brightest inner regions of the galaxy.

M31 has two small companion galaxies visible in small telescopes. One Moon diameter south of M31 lies **M32**, looking like a fuzzy 8th-magnitude star. The other, **M110**, lies twice as far to the north and is more difficult to spot, being larger and more diffuse.

Seen from outside, our own Galaxy would look much like M31. Of the two, M31 seems to be slightly the larger, with a diameter of about 150,000 light years and containing 400,000 million stars, twice the number in our Milky Way. Do any of those stars have planets on which inquisitive life forms are looking back at us?

Our Galaxy and M31 are the two largest members of a cluster of about three dozen galaxies called the Local Group. To find another member of the Local Group, look over the border from Andromeda into Triangulum. There lies **M33**, another spiral galaxy, somewhat farther off than M31. It is presented face-on to us, and covers a larger area of sky than the full Moon, but it is notoriously difficult to detect because its light is so pale. Dark skies are essential to see M33, and binoculars are better than telescopes because they condense its thinly spread light. Sometimes observers overlook M33 because it is much larger than they expect.

Pegasus

Pegasus represents the winged horse that was born from the blood of Medusa when she was beheaded by Perseus. Only the front quarters of the horse are depicted in the sky. Pegasus is sometimes identified as the horse of Perseus, but in fact it was another hero, Bellerophon, who rode Pegasus. The constellation is the seventh-largest in the sky, but contains no outstandingly bright stars. Its most distinctive feature is the **Great Square of Pegasus** whose corners are marked by four stars of 2nd and 3rd magnitudes, although only three of the stars actually belong to Pegasus. The fourth star is Alpha Andromedae, which on some old maps was called Delta Pegasi.

Pegasus is upside-down in the sky. The Great Square represents his body and the crooked line of stars leading from the bottom right of the Square to Epsilon (ε) Pegasi marks his neck and head. The two lines of stars from the upper right of the Square are his forelegs.

The Square of Pegasus is a large feature, with a diameter of over 15°, or about the width of two outstretched fists, enough to contain a line of more than 30 Moons side by side. Yet within this large area of sky, only a handful of stars are visible to the naked eye. How many stars can you count here?

The star at the top right of the Square is **Beta** (β) **Pegasi**, an immense red giant approximately 100 times the diameter of the Sun. It varies irregularly in size and brightness between magnitudes 2.3 and 2.7. Compare it with the other stars of the Square, Alpha (α) Pegasi, magnitude 2.5, and Gamma (γ) Pegasi, magnitude 2.8.

About halfway down the right hand side of the Square lies an unassuming 5th-magnitude star. This is **51 Pegasi**, which achieved lasting fame in 1995 when it became the first star beyond the Sun discovered to have a planet orbiting it. The star itself, which lies 50 light years away, is similar to the Sun. The planet has a mass about half that of Jupiter but swings around 51 Pegasi in only 4 days, a remarkably short "year" which means that it must be much closer to the star than any of the planets in our own Solar System. Other examples of stars with planets have since been confirmed, including the first multiple-planet system around Upsilon (υ) Andromedae (*see facing page*).

The brightest star in the constellation is **Epsilon** (ε) **Pegasi**, otherwise known as Enif, from the Arabic meaning 'nose'. It is a brilliant orange supergiant nearly 700 light years away, appearing of magnitude 2.4. Binoculars or small telescopes show it to have a wide companion star of 8th magnitude.

Above and to the right of Epsilon Pegasi is the greatest treasure of this constellation, the globular cluster **M15**. It is just too faint to be seen with the naked eye under normal conditions, but shows up clearly in binoculars as a rounded, fuzzy patch of light. A 6th-magnitude star lies nearby, acting as a guide to its location.

Small telescopes show that the cluster has a bright core, where its stars crowd together most densely, although small apertures will be unable to resolve individual stars. The rest of the cluster is like a misty halo around this bright centre. M15 lies 30,000 light years away.

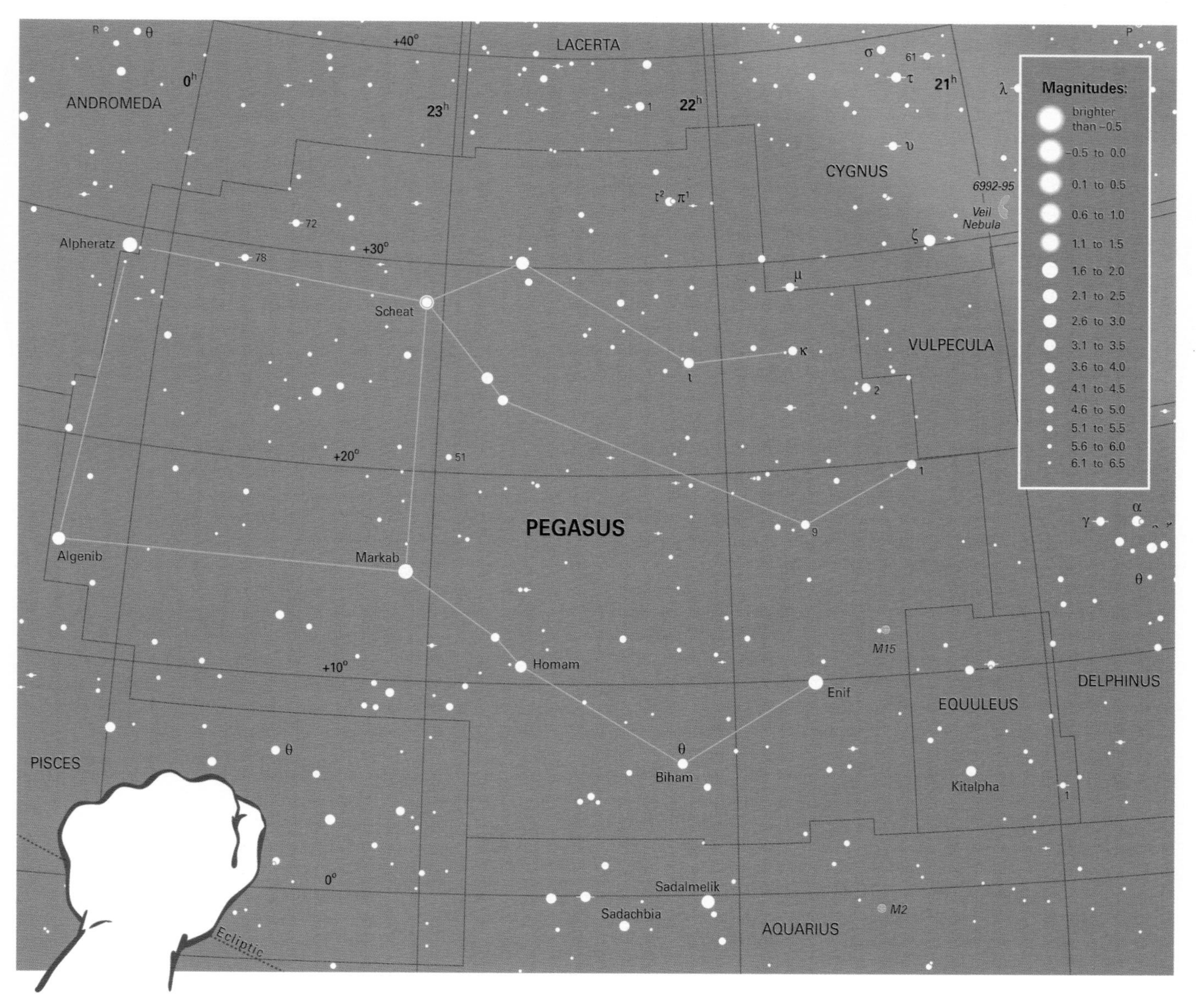

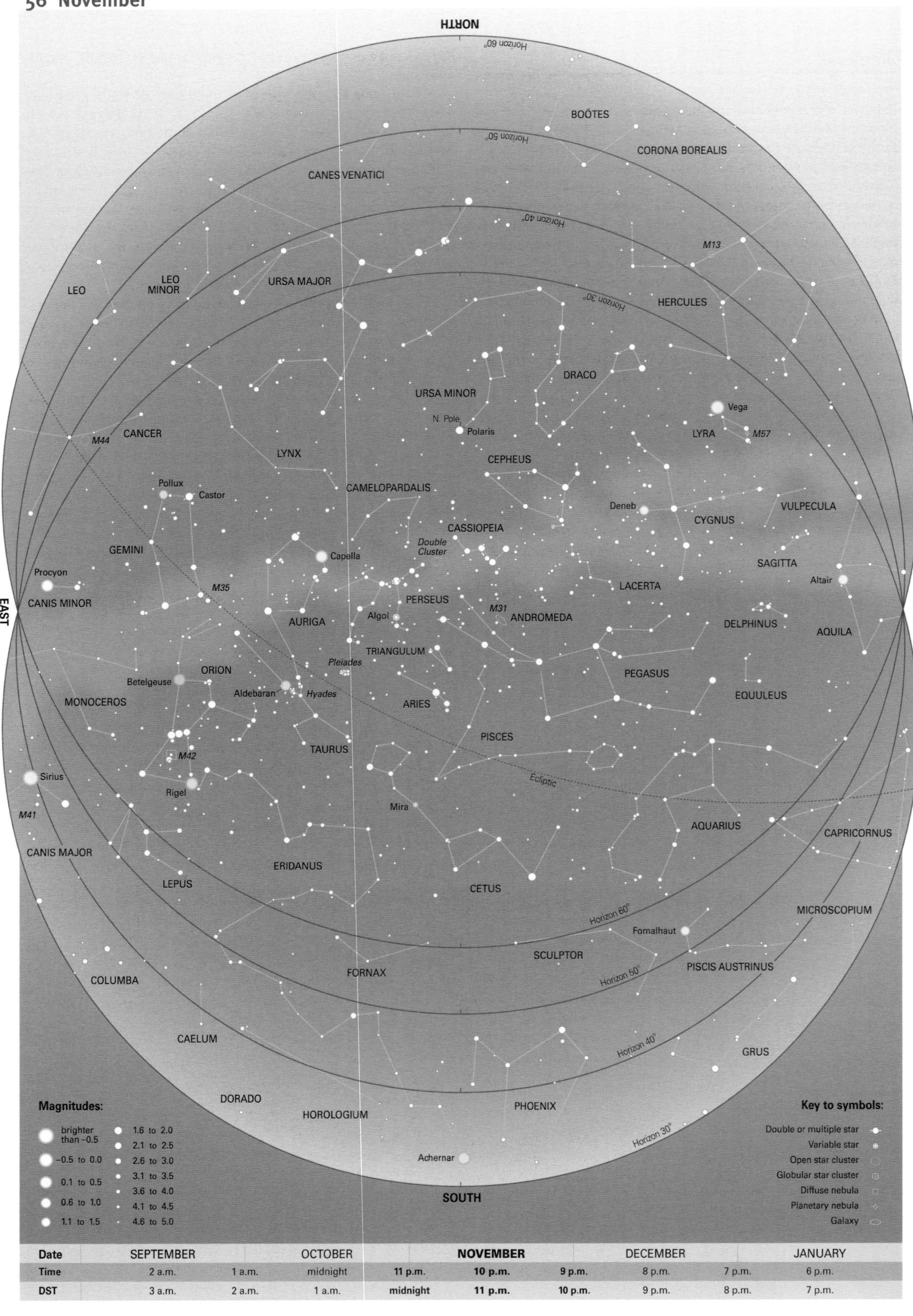

Date	SEPTEMBER		OCTOBER		**NOVEMBER**		DECEMBER		JANUARY
Time	2 a.m.	1 a.m.	midnight	**11 p.m.**	**10 p.m.**	**9 p.m.**	8 p.m.	7 p.m.	6 p.m.
DST	3 a.m.	2 a.m.	1 a.m.	**midnight**	**11 p.m.**	**10 p.m.**	9 p.m.	8 p.m.	7 p.m.

November

Key Stars

The Milky Way arches overhead, passing through Cassiopeia and Perseus. The Summer Triangle reluctantly departs in the west, but the Square of Pegasus is still high in the south-west. Due south is the barren area of Pisces, Cetus and Eridanus, with no prominent stars, although the 1st-magnitude star Achernar just peeps above the southern horizon as seen from 30° north. Aldebaran, the red eye of Taurus the Bull, and yellow Capella stand prominently in the south-east, followed by Gemini and Orion, signalling the approach of winter.

The planets this month

Venus

2007 Prominent in the eastern morning sky at mag. –4.3.
2008 In the south-west evening sky at mag. –4.1. Ends the month near fainter Jupiter.
2009 In the eastern dawn sky, becoming lost in twilight by mid-month.
2010 Moves rapidly away from the Sun into the south-east dawn sky during the first week, brightening to a dazzling mag. –4.7 at month's end.
2011 In the south-west evening sky at mag. –3.9, but too low to be seen well from more northerly latitudes.

Mars

2007 Stationary in Gemini and brightening rapidly to mag. –1.3 during the month.
2008 Too close to the Sun for observation throughout the month.
2009 A morning object in Cancer, brightening to mag. 0.1.
2010 Lost in the south-western evening twilight throughout the month.
2011 A first-magnitude morning object in Leo, passing Regulus in mid-month.

Jupiter

2007 Low in the south-west, becoming lost in evening twilight during the month.
2008 In the south-west evening sky at mag. –2.0, ending the month above brilliant Venus.
2009 A prominent evening object in Capricornus at mag. –2.4.
2010 A prominent evening object, stationary on the borders of Pisces and Aquarius at mag. –2.6.
2011 Visible all night in southern Aries at a prominent mag. –2.9.

Saturn

2007 A morning object, under the body of Leo at mag. 0.8.
2008 A first-magnitude morning object under the hind quarters of Leo.
2009 In the south-eastern dawn sky at mag. 1.1.
2010 In the south-eastern dawn sky among the stars of Virgo at mag. 0.9.
2011 Emerges into the south-east dawn twilight in the second half of the month at mag. 0.8.

Eclipses

Sun

2011 November 25. Partial solar eclipse, maximum 90%, visible from Antarctica, Tasmania, southern New Zealand and southern South Africa.

http://sunearth.gsfc.nasa.gov/eclipse/SEcat/SEdecade2011.html

November meteors

Two meteor showers can be seen in November: the Taurids and the Leonids. The **Taurids** have an extended maximum that lasts for several days either side of November 3, when about ten meteors an hour may be seen coming from the region near the Hyades and Pleiades clusters. There are two radiants to this stream, one 8° north of the other. Taurid meteors are debris from Comet Encke, the comet with the shortest known orbital period, 3.3 years. Taurids are slow-moving and bright, which makes the display more impressive than their low numbers might suggest.

Altogether more spasmodic is the **Leonid** shower, which radiates from near Gamma (γ) Leonis, peaking on November 17. Leonids are swift, stabbing meteors, often flaring at the end of their paths. Many leave persistent trains. Since Leo does not rise until late, observations have to be made after midnight. The Leonids are associated with Comet Tempel–Tuttle, which orbits the Sun every 33 years. Rates are usually modest, no more than ten an hour. But spectacular Leonid storms can occur at 33-year intervals, when Comet Tempel–Tuttle returns to the inner region of the Solar System and the Earth encounters a dense swarm of cometary dust. In 1966 observers in the United States saw Leonid meteors falling from the sky like rain, reaching a peak of thousands a minute. A strong display occurred in 1998, when the Leonids' parent comet last returned, and was followed by rates of a few thousand an hour in 1999, 2001 and 2002.

Perseus

Perseus, a son of Zeus, is the hero in the most enduring of Greek myths, that of Perseus and Andromeda. His adventures began when he was sent to kill Medusa, one of the three Gorgons, winged monsters with snakes for hair who were so terrible to behold that anyone who looked at them was turned to stone – literally petrified. Perseus sneaked up on Medusa while she was sleeping, looking only at her reflection in his shield so that he escaped being petrified. He decapitated Medusa and flew off with her head in a bag. While flying home, he saw the beautiful maiden Andromeda chained to a rock as a sacrifice to a sea monster. Perseus swooped down, killed the monster and claimed Andromeda as his bride.

In the sky, Perseus is depicted holding aloft the head of Medusa, marked by **Beta** (β) **Persei**, known by its Arabic name of **Algol**, meaning 'ghoul' or 'demon'. The name is certainly appropriate, for Algol is a variable star that appears to 'wink' like an evil eye at regular intervals, from magnitude 2.1 to 3.4. Algol's variations in light were first recognized by the Italian astronomer Geminiano Montanari in 1667, and the periodicity of the variations was first measured by the English amateur astronomer John

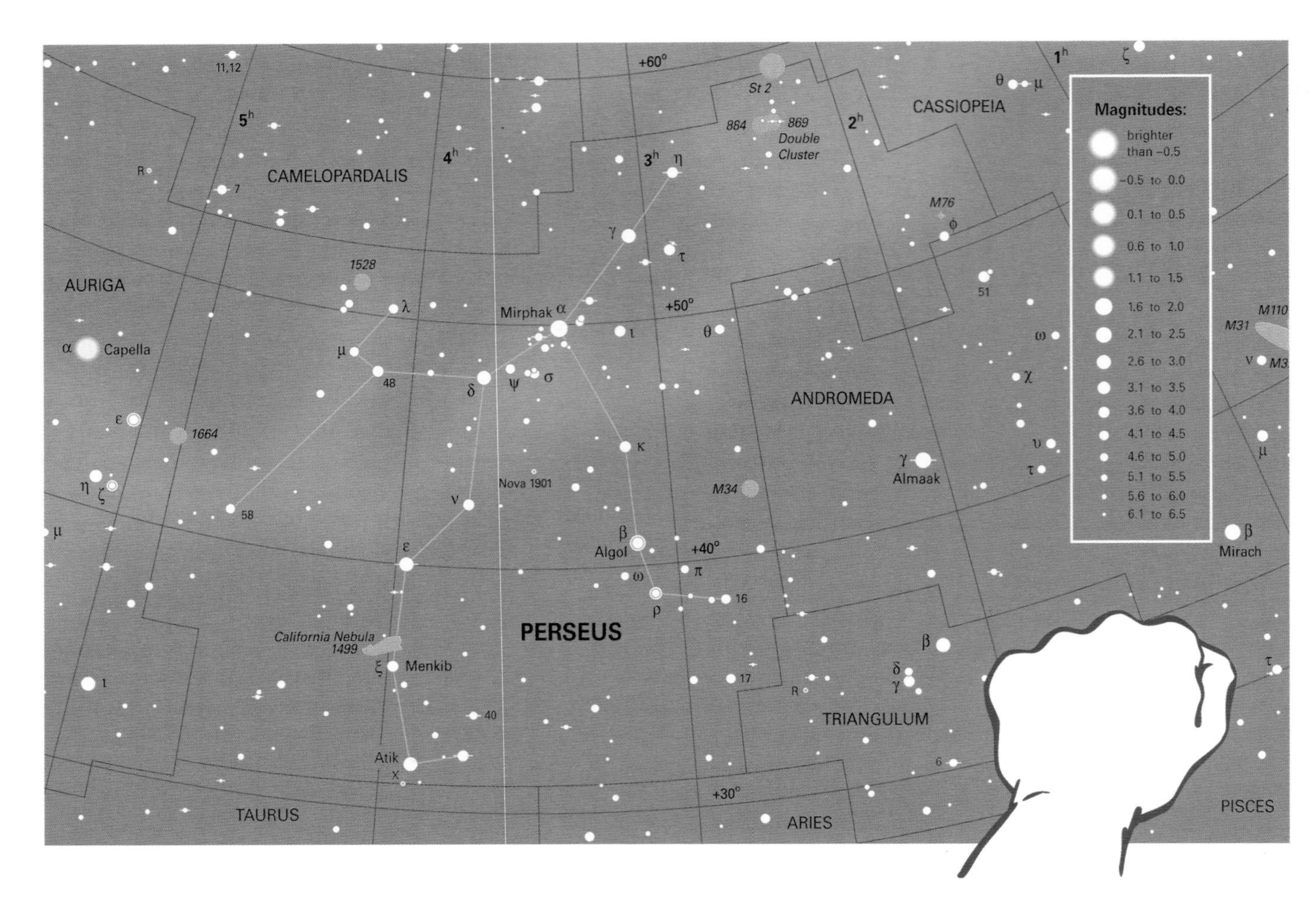

Goodricke in 1783. Goodricke also suggested, correctly, that the variations were caused by eclipses.

Algol is an eclipsing binary, consisting of two stars that regularly pass in front of each other as seen from Earth. The stars are too close together to be distinguished individually in a telescope, but analysis of the light from Algol reveals that the brighter of the pair, Algol A, is a hot star 100 times as luminous as the Sun. Its companion, Algol B, is a larger but fainter orange star that covers about 80 per cent of Algol A during the eclipses. There is also a third star in the system, Algol C, but this does not participate in the eclipses.

Algol's eclipses occur every 2 days 20 hours 49 minutes. In a mere five hours, Algol fades to one-third of its usual brightness, from magnitude 2.1 to 3.4, then returns to magnitude 2.1 in another five hours. These variations are easily followed with the naked eye. Compare Algol with Beta (β) Cassiopeiae, magnitude 2.3; Delta (δ) Cassiopeiae, mag. 2.7; Epsilon (ε) Persei, mag. 2.9; Delta (δ) Persei, mag. 3.0; Alpha (α) Trianguli, mag. 3.4; and Kappa (κ) Persei, mag. 3.8. You should end up with a light curve looking like the one shown below. A secondary minimum occurs when Algol A eclipses Algol B, but the fading is so slight that it is undetectable to the eye. Incidentally, do not compare Algol with

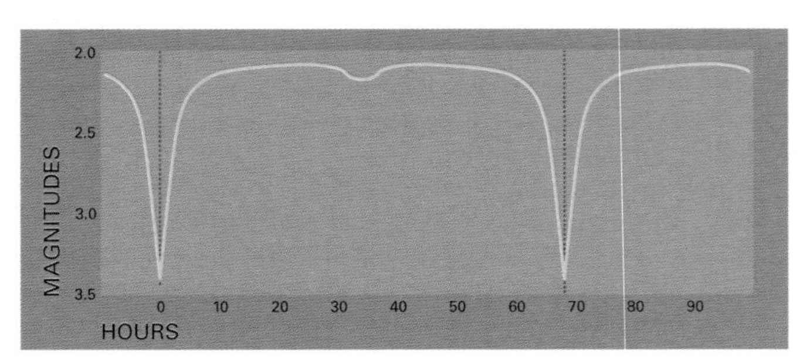

Graph of brightness changes of the variable star Algol, in Perseus.

Rho (ρ) **Persei**, which is a red giant that fluctuates somewhat irregularly between magnitudes 3.3 and 4.0 every month or two.

Perseus lies in a sparkling part of the Milky Way, rich for sweeping with binoculars. Look at the region around the constellation's brightest star, **Alpha** (α) **Persei**, magnitude 1.8. Alpha Persei is a yellow-white supergiant 5000 times more luminous than the Sun, surrounded by a sprinkling of bright stars that form a loose cluster 600 light years away. Note also the cluster **M34** on the border with Andromeda, an excellent object for binoculars and small telescopes, consisting of several dozen stars of 8th magnitude and fainter, covering a similar area of sky to the full Moon. M34 lies 1400 light years away.

But the real showpiece of Perseus is the **Double Cluster**, a twin star cluster on the border with Cassiopeia, marking the hand of Perseus holding aloft his sword. To the naked eye, the clusters resemble a denser knot in the Milky Way. Binoculars and small telescopes show this area to be sown with stars, one of the finest sights in the sky for small instruments. Note an arc of stars that leads northwards to a large, scattered cluster on the border with Cassiopeia, known as **Stock 2**.

Each half of the Double Cluster covers a greater area of sky than the full Moon. However, the two clusters are not identical twins. **NGC 869**, also known as **h Persei**, is the more star-packed of the two. The other cluster, **NGC 884**, also known as **Chi** (χ) **Persei**, contains some red giant stars, which NGC 869 does not, probably because NGC 884 is older so that the stars in it have evolved more. Both clusters lie in a spiral arm of our Galaxy. NGC 884 is slightly the more distant at 7500 light years, while NGC 869 is about 300 light years closer. To be visible over such distances the brightest stars in each cluster must be intensely luminous supergiants, similar to Rigel and Betelgeuse in Orion. If the Double Cluster were as close as the stars of Orion, it would appear as spectacularly large and bright to the naked eye as we see it through binoculars.

Auriga

Next to Perseus in the sky is the hexagonal shape of Auriga, representing a man driving a chariot. By comparison with the flamboyant figure of Perseus, the identity of Auriga is somewhat shadowy. He is usually said to be Erichthonius, a lame king of Athens who invented the four-horse chariot. But another legend identifies Auriga with Myrtilus, charioteer of King Oenomaus and a son of Hermes.

There is no mistaking the constellation's brightest star, Alpha (α) Aurigae, better known as **Capella**. At magnitude 0.1 Capella is the sixth-brightest star in the sky. Its name comes from the Latin meaning 'little she-goat', for the charioteer has traditionally been depicted carrying a goat on his left shoulder. Capella is actually a pair of yellow giants forming a spectroscopic binary, 42 light years away. It is the most northerly of all 1st-magnitude stars.

The stars Eta (η) Aurigae and Zeta (ζ) Aurigae are known as the Kids, the goat's offspring, carried on the charioteer's arm. **Zeta Aurigae** is one of two extraordinary eclipsing binary stars in this constellation. It consists of an orange giant some 150 times larger than the Sun, orbited by a blue-white star similar to Regulus. Although this blue-white star is smaller than the much brighter giant, it is still about four times the diameter of the Sun. The luminosities of the two stars are 700 and 140 Suns, respectively. Normally Zeta Aurigae shines at magnitude 3.7, but every 2 years 8 months the smaller star is eclipsed by the red giant and the brightness falls by a third, to magnitude 4.0. The eclipses last six weeks, after which the star returns to its normal brightness.

Even more extraordinary is **Epsilon** (ε) **Aurigae**, which has the longest known period of any eclipsing binary, 27 years. The main star is an intensely luminous white supergiant, one of the most powerful stars known, shining with the light of over 100,000 Suns and large enough to contain the orbit of the Earth. It lies at least 2000 light years away. Normally it appears at magnitude 3.0, but every 27 years it is partially eclipsed by a mysterious dark companion. Over a period of four months the star's brightness gradually halves, to magnitude 3.8, where it remains for 14 months before returning to normal.

The nature of Epsilon Aurigae's dark companion was a mystery to astronomers for years. It is now thought to consist of a close pair of stars surrounded by a dark disk of dust that actually causes the eclipses. Epsilon Aurigae was last eclipsed in 1982 to 1984, and the next eclipse will begin in late 2009.

Auriga is notable for an impressive trio of star clusters, **M36**, **M37** and **M38**, all three being visible in the same field of view through wide-angle binoculars. In binoculars they appear as fuzzy patches, but small telescopes resolve them into individual stars. Each cluster has its own distinct character.

M36 is the smallest and most condensed of the trio, consisting of 60 or so stars of 9th magnitude and fainter, lying 3900 light years away. In binoculars it appears the most prominent of the Auriga clusters. The largest and richest of the Auriga clusters is M37, containing 150 or so stars 4200 light years away. At its centre is a brighter orange star. The most scattered of the clusters is M38, containing about 100 faint stars, 3900 light years away.

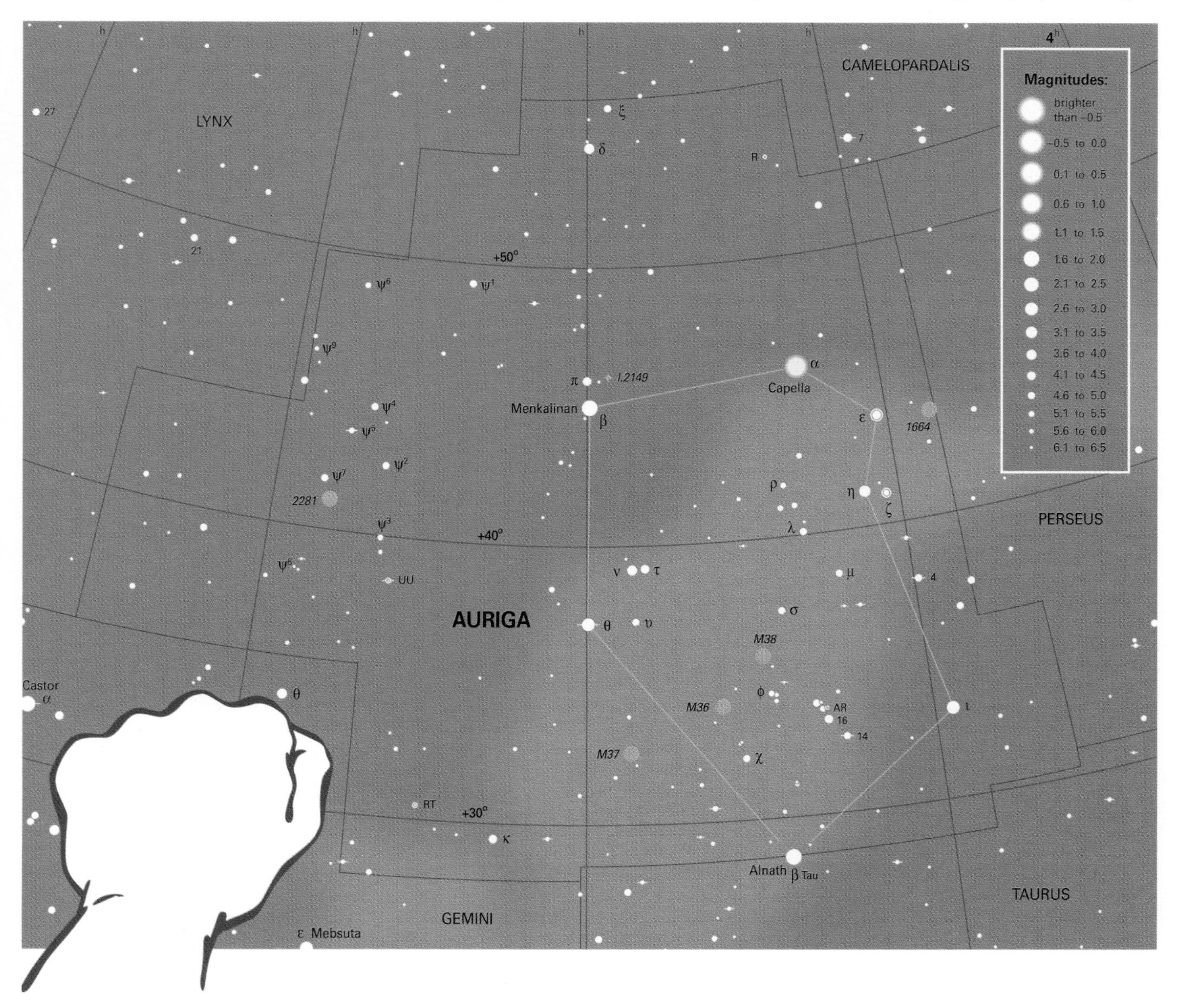

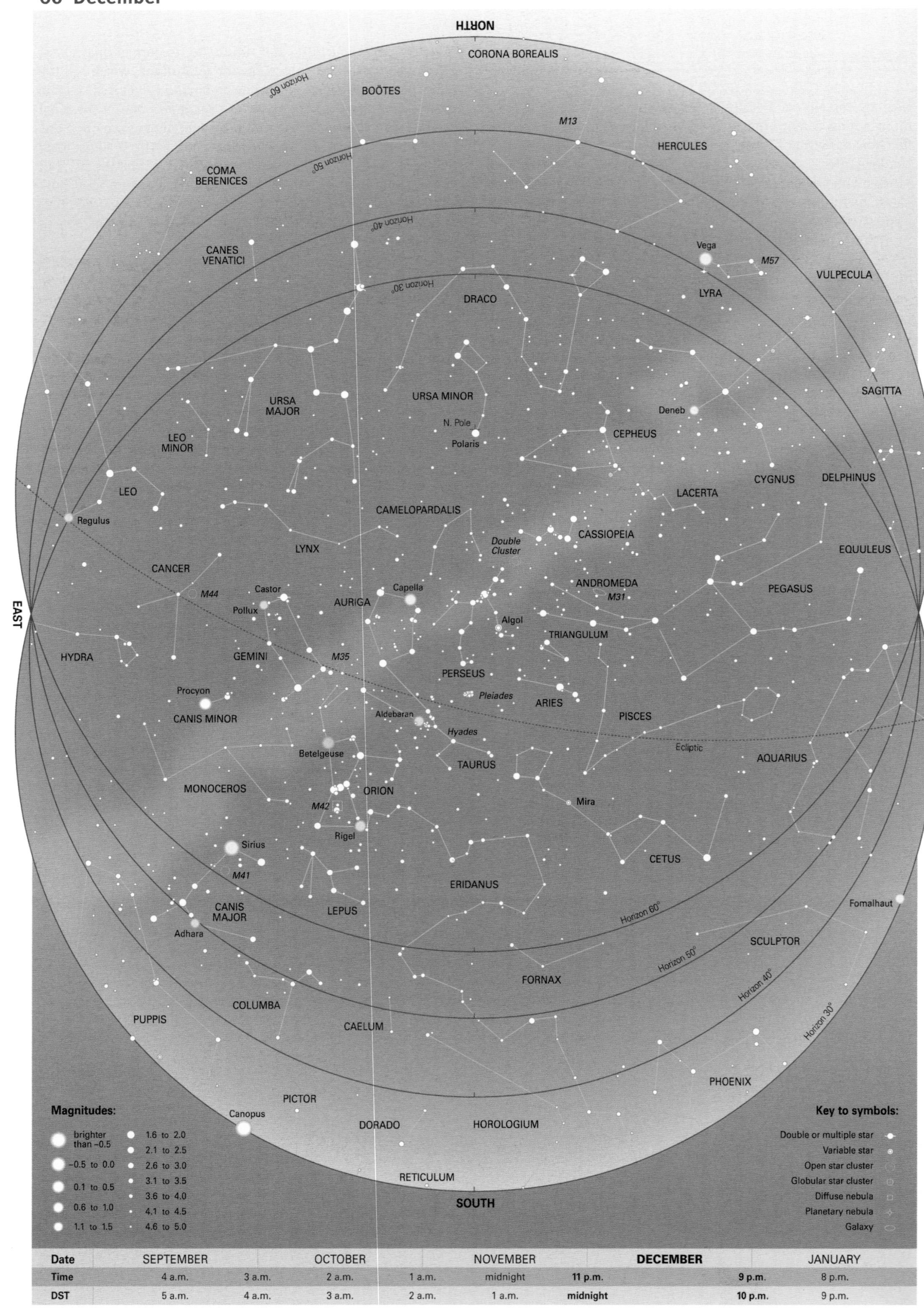

Date	SEPTEMBER		OCTOBER		NOVEMBER		**DECEMBER**		JANUARY
Time	4 a.m.	3 a.m.	2 a.m.	1 a.m.	midnight	**11 p.m.**		**9 p.m.**	8 p.m.
DST	5 a.m.	4 a.m.	3 a.m.	2 a.m.	1 a.m.	**midnight**		**10 p.m.**	9 p.m.

December

Key Stars

The misty-looking Pleiades cluster is due south at 10 p.m. in mid-month. Above it is Perseus, flanked by W-shaped Cassiopeia and boxy Auriga. Following the Pleiades across the sky is the V-shaped Hyades group with ruddy Aldebaran at one tip. Orion marches grandly upwards from the south-east followed by his dogs, brilliant Procyon and Sirius. Between Orion and the pole glitters Capella. High in the east are Castor and Pollux, the twin stars of Gemini. Below them, on the eastern horizon, the head of Leo is rising. In the south-west is the Square of Pegasus, making a diamond shape as it stands on one corner. Vega lingers on the north-western horizon, for observers above latitude 45° north.

The planets this month

Venus

2007 In the south-east dawn sky at mag. –4.1.
2008 Prominent in the south-west evening sky at mag. –4.2, passing fainter Jupiter at the start of the month.
2009 Too close to the Sun for observation throughout the month.
2010 Blazes in the south-east dawn sky at mag. –4.6 throughout the month.
2011 Low in the south-west evening sky at mag. –4.0.

Mars

2007 Visible all night in Gemini. Opposition (due south at midnight) is on December 24 at mag. –1.6, 89 million km (55 million miles) from Earth.
2008 At conjunction (behind the Sun) on December 5 and invisible throughout the month.
2009 Stationary between Leo and Cancer, brightening rapidly from mag. 0.0 to –0.7 during the month.
2010 Too close to the Sun to be seen throughout the month.
2011 A morning object under the body of Leo, brightening to mag. 0.2 by month's end.

Jupiter

2007 Too close to the Sun to be seen throughout the month. At conjunction (behind the Sun) on December 23.
2008 Low in the south-west evening sky at mag. –1.9. Starts the month to the right of brilliant Venus and ends it to the right of mag. –0.7 Mercury.
2009 In the south-west evening sky in Capricornus at mag. –2.2.
2010 Prominent in the south-west evening sky on the Pisces–Aquarius border at mag. –2.4.
2011 Prominent for most of the night on the Aries–Pisces border at mag. –2.7.

Saturn

2007 Stationary under the body of Leo at mag. 0.7.
2008 A first-magnitude morning object under the hind quarters of Leo.
2009 In the south-east morning sky in Virgo at mag. 1.0.
2010 In the south-east morning sky in Virgo at mag. 0.8.
2011 In the south-east dawn sky in Virgo at mag. 0.7.

Eclipses

Moon

2010 December 21. Total lunar eclipse, visible from north-west Europe, the Americas and north-east Asia. Totality starts 7h 40m GMT, ends 8h 53m GMT.
http://sunearth.gsfc.nasa.gov/eclipse/lunar.html

2011 December 10. Total lunar eclipse, visible from north-west North America, Australasia, Asia, eastern Europe. Totality starts 14h 06m, ends 14h 58m.
http://sunearth.gsfc.nasa.gov/eclipse/LEcat/LEdecade2011.html

December meteors

The **Geminids** are one of the brightest and richest showers of the year. At their peak, around December 13–14, as many as 100 meteors an hour may be seen radiating from near Castor. The radiant is well placed for observation throughout the night. Geminid meteors are bright, like the Perseids, but unlike the Perseids they rarely leave trains. Activity falls off sharply after maximum. The Geminids are unique in that their parent body is not a comet, but an asteroid. This asteroid, which has an orbital period of 1.4 years, was discovered in 1983 and was named Phaethon; it is probably a 'dead' comet.

Taurus

As the year draws to a close, the familiar figure of Taurus, the Bull, snorts the night air in the south. His face is marked by the V-shaped cluster of stars called the Hyades, his glinting red eye is the bright star Aldebaran and his long horns are tipped by the stars Zeta (ζ) and Beta (β) Tauri. In the past, Beta Tauri was regarded as being common to both Auriga and Taurus and on old maps it was given the alternative designation of Gamma (γ) Aurigae, but now it is the exclusive property of Taurus. Appropriately enough, its Arabic name, Alnath, means 'the butting one'.

Only the top half of the Bull is depicted in the sky. This might be explained by one legend, which identifies Taurus with the bull into which Zeus changed himself to seduce Princess Europa of Tyre. He carried her on his back through the ocean, so perhaps the rest of the Bull is immersed beneath the waves. Another identification of Taurus is as the Cretan bull tamed by Hercules as one of his twelve labours. Alternatively, Taurus could represent the Minotaur, the monster half-bull, half-man slain by Theseus.

The brightest star in the constellation, Alpha (α) Tauri, better known as **Aldebaran**, has a noticeably fiery glow, like a red-hot coal, for it is an orange giant star, 25 times the diameter of the Sun and 150 times as bright. The name Aldebaran comes from the Arabic meaning 'the follower', from the fact that it follows the Hyades and Pleiades star clusters across the sky. Aldebaran lies 65 light years away and appears of magnitude 0.9.

Aldebaran looks as though it is a member of the **Hyades** star cluster, but it is not. It is actually a foreground object at less than half the distance, superimposed on the Hyades by chance.

The Hyades and Pleiades clusters are two of the major tourist attractions in the sky. In mythology, the Hyades and Pleiades were sisters, the daughters of Atlas, but Aethra was mother of the Hyades and Pleione was mother of the Pleiades. The Greeks told that the Hyades nursed the infant Dionysus, a son of Zeus, and were rewarded with a place in the heavens. The name Hyades means 'the rainy ones', because they were associated with bad weather.

Legend says that there were five Hyades, but at least a dozen stars are visible to the naked eye. The Hyades cluster spans 5° of sky, so large that only binoculars can encompass it. As you scan among the numerous stars that spring into view here, note several bright, wide doubles including **Theta** (θ) **Tauri** and **Sigma** (σ) **Tauri**. At magnitude 3.4, Theta-2 (θ^2) Tauri is the brightest star in the cluster, followed closely by Epsilon (ε) Tauri, magnitude 3.5, and Gamma (γ) Tauri, magnitude 3.6.

Hundreds of stars belong to the Hyades, all of them moving through space together towards a point in the sky near Betelgeuse. Such a moving cluster is of particular importance to astronomers because the cluster's distance can be accurately derived from the movements of its stars. Measured in this way, the Hyades turns out to be 150 light years away, making it the nearest star cluster to us. The distance of the Hyades is the first stepping stone in our scale of the Universe.

But even the Hyades is overshadowed by its near neighbour the **Pleiades**, popularly known as the Seven Sisters, the most famous of all star clusters. The Pleiades hover like a swarm of flies over the Bull's back. To a casual glance, the Pleiades appear as a misty patch, but good eyesight reveals six or seven individual stars and some people with exceptional eyesight can see ten or more. The misty effect is produced by the dozens of fainter stars that are just beyond visibility with the naked eye. Without doubt, the Pleiades cluster is the finest binocular sight in astronomy.

According to myth the Pleiades were seven nymphs, followers of Artemis, the huntress. Orion, smitten by their beauty, began to pursue them. To save them from his advances, Artemis placed them among the stars. And in the heavens, Orion still forlornly chases the Pleiades across the sky.

Unlike the Hyades cluster, none of whose stars are named, the stars of the Pleiades bear the names of all seven sisters: Alcyone, Asterope, Celaeno, Electra, Maia, Merope and Taygeta. Their father and mother, Atlas and Pleione, are also present. There is even a legend to explain why only six Pleiades, rather than seven, are easily visible. Electra is said to have hidden her face so as not to witness the ruins of Troy, the city founded by her son Dardanus. But some say that the 'missing' Pleiad is Merope, ashamed because she was the only one of the sisters to wed a mortal. In fact, the naming of the Pleiades has not followed either legend because the faintest of the named stars in the cluster are Asterope and Celaeno.

In addition to the named stars, binoculars bring into view a glittering array of dozens more, fitting snugly into the field of view. At least 100 stars are members of the cluster, lying 380 light years away, two and a half times as far as the Hyades. The main stars of the Pleiades form a shape resembling a squashed version of the Plough or Big Dipper. An interesting demonstration of comparative sizes in the sky is the fact that the full Moon could fit into the bowl of the Pleiades 'dipper' without obscuring either Alcyone or Taygeta. The whole cluster extends across more than two Moon widths.

The brightest member of the Pleiades is **Alcyone**, also known as Eta (η) Tauri, magnitude 2.9, a blue-white giant 800 times more luminous than the Sun. **Pleione**, also known as **BU Tauri**, is an interesting variable of the type known as a shell star. It rotates so quickly, about 100 times faster than our Sun, that it is unstable and throws off shells of gas, causing it to vary gradually in brightness; currently it is around 5th magnitude. Perhaps Pleione was much brighter in the past and is really the 'missing' Pleiad.

The Pleiades cluster is relatively young, with an estimated age of 50 million years, compared with 600 million years for the

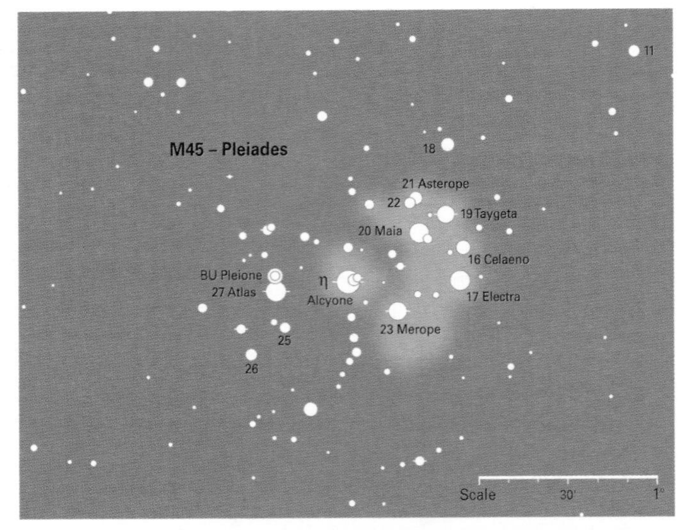

Pleiades

Alcyone	mag. 2.9
Atlas	mag. 3.6
Electra	mag. 3.7
Maia	mag. 3.9
Merope	mag. 4.1
Taygeta	mag. 4.3
Pleione	mag. 4.8–5.5 (variable)
Celaeno	mag. 5.5
18 Tauri	mag. 5.7
Asterope	mag. 5.8
22 Tauri	mag. 6.4

Hyades. But the brightest members of the Pleiades, which stud the binocular field like priceless blue-white diamonds, are no more than a few million years old. Long-exposure photographs show that the Pleiades stars are surrounded by traces of gas and dust, but this nebulosity is not visible through amateur instruments under any but the most exceptional conditions. The nebulosity was once thought to be the remnants of the cloud from which the stars formed, but more recent research suggests it is an unrelated cloud into which the cluster has since drifted by chance.

Taurus is blessed with exceptional objects, for in addition to the Hyades and Pleiades clusters it contains the **Crab Nebula**, perhaps the most-studied object outside the Solar System. Astronomers also know it as **M1**, the first in the list of nebulous objects compiled by the Frenchman Charles Messier.

The Crab Nebula is the remains of a star much more massive than the Sun that ended its life as a supernova, a cataclysmic nuclear explosion. Astronomers in China saw the star flare up in July 1054. At its brightest it reached magnitude −4, similar to Venus, and was visible in daylight for three weeks. It finally faded from naked-eye view after 21 months, in April 1056. There are no European records of the explosion, probably because astronomy was almost non-existent in Europe at that time.

During its explosion the Crab supernova must have blazed with the brilliance of 500 million Suns. Twenty-five such supernovae could equal the entire light output of our Galaxy. The outer layers of the shattered star, splashed into space at high speed by

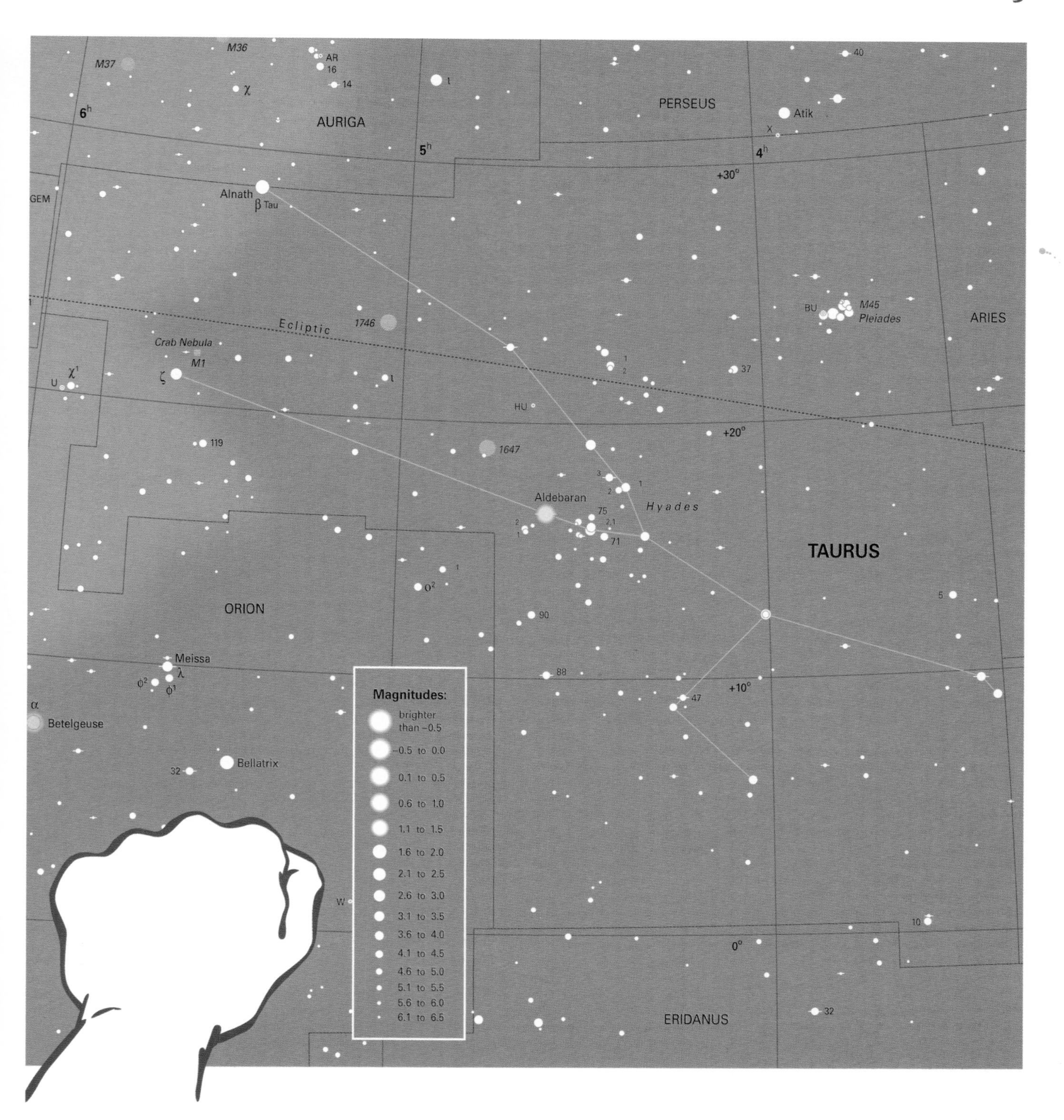

the explosion, now form the Crab Nebula, a name due to the 19th-century Irish astronomer Lord Rosse who thought that its shape resembled a hermit crab.

The Crab Nebula lies between the Bull's horns, just over 1° (two Moon diameters) from Zeta (ζ) Tauri, and 6500 light years from Earth. Under good conditions the Crab Nebula can be glimpsed in binoculars, but for most observers a telescope will be necessary to show its elongated misty blur, shaped rather like the flame of a candle. It measures 5 minutes of arc across, one-sixth the Moon's width but six times larger than the disk of Jupiter, so look for an object midway in size between the Moon and a planet. Alas, it is far less spectacular in a telescope than it appears in photographs, but no observer's log is complete without a sighting of the Crab.

At the centre of the Crab Nebula is one of the exceedingly small and dense objects known as neutron stars, the crushed core of the original star that exploded. A neutron star contains as much mass as the Sun, squeezed into a ball perhaps 20 km (12 miles) across. Being so small, the neutron star can spin very quickly – 30 times a second in the case of the star in the Crab. Each time it spins, the neutron star emits a flash of light, radio waves and X-rays, like a lighthouse beam. Such a flashing neutron star is termed a pulsar. But, at 16th magnitude, the Crab pulsar is within reach of large telescopes only.

It is a startling fact that, without supernovae, we would not be here. In a supernova, nuclear reactions convert the original hydrogen and helium into all the chemical elements of nature, which are then scattered into space, later to be collected up into new stars, planets and, perhaps, life. When you look at the Crab Nebula, recall that the atoms of your own body were produced in such a way, in the supernova explosion of a star that lived and died long before the Earth was born.

Index

Further reading

Those wishing to progress to more advanced observing will find descriptions of interesting objects for telescopes up to 200 mm aperture, along with maps of every constellation in the sky, in *Collins Pocket Guide to Stars and Planets* by Ian Ridpath and Wil Tirion (published in the US as the *Princeton Field Guide to Stars and Planets*). For more on the myths of the constellations see *Star Tales* by Ian Ridpath (Lutterworth UK, Universe US).

A complete atlas of the sky down to magnitude 6.5 is the *Cambridge Star Atlas* by Wil Tirion (Cambridge University Press).

http://www.ianridpath.com/
http://www.wil-tirion.com/